BOOKS BY WILLIAM H. AMOS

Life of the Seashore
Life of the Pond
The Infinite River

THE
INFINITE
RIVER

*I saw a distant river by moonlight, making
no noise, yet flowing, as by day, still to
the sea, like melted silver reflecting the
moonlight. Far away it lay encircling the
earth . . . . There is a certain glory attends
on water by night. By it the heavens are
related to the earth, undistinguishable from
a sky beneath you.*

THOREAU

William H. Amos

THE

INFINITE

RIVER

A Biologist's Vision of the

World of Water

RANDOM HOUSE

NEW YORK

For my family—
and for all who love this earth

PREFACE

This world, after all our science
and sciences, is still a miracle;
wonderful, inscrutable, magical
and more, to whosoever will think
of it.

THOMAS CARLYLE

W H Y is this account only a vision? Partly because it
is composed of events and features of many major river
systems; partly because no one person could hope to be in
all the places described; but most important, because these
conditions did in fact exist not very long ago—and many no
longer do.

No major river system remains unpolluted in North
America or, for that matter, anywhere else. The Ganges,
Thames, Nile, Yangtze, Danube, Rhine, and Amazon, as
well as our own more familiar rivers, have their man-in-
duced illnesses. The atmosphere that gives birth to a river
is filled with foreign gases and particles. The seas into
which the rivers flow contain dissolved substances entirely
new in their three- or four-billion-year history.

Quite properly, we are apprehensive of the spread of disease through our own populations. We make every effort
to combat such debilitating events in our lives, yet man
himself has become a pathogen within the global ecosystem,

the world environment which gave rise to him only recently in its long and complex history.

In the United States—as elsewhere—we have impounded
rivers, let them fill with silt, and sickened them with wastes
beyond accounting. We still have within our technologies
the means of helping rivers to regain their health, for each
has built into its dynamics the ability to recover despite
grievous wrongs—up to a doomsday point. Surely there is
a time at which the biosphere, a shell of life largely dependent upon water in motion, can be poisoned beyond recovery.

I have chosen not to write a series of alarms or case his-
tories of an increasingly despoiled world (these abound and
are available to all), but a narrative which touches a few of
the intricacies and interrelationships of the world of water.
The book suggests what once was, what today exists only
in part, and what some despair of seeing again. It will serve
well if it provides a brief comparison between a past equilibrium and a present imbalance. If it induces further con-
cern in either student or citizen, we shall all benefit. And if
it does neither, perhaps the book will say something to the
nonobserver about whom Wordsworth wrote,

> *A primrose by a river's brim*
> *A yellow primrose was to him,*
> *And it was nothing more.*

The decision to write a book in this manner was mine
alone, having evolved as a challenge and an obligation after
following many rivers from their upland sources to a final
merging with the sea, a pursuit that has been both exhilarat-
ing and depressing. Opportunities for observation were
provided, either specifically or during periods of study and
research, by many laboratories and institutions during the
last two decades, including the University of Delaware
Marine Laboratories, St. Andrew's School, and the Marine

Biological Laboratory at Woods Hole. Certain aspects of the work were generously assisted by the Lalor Foundation and the Bredin Foundation.

For those who find it perplexing not to know the identity and location of this river, I can help only to this extent: it is a composite of several major river and estuarine systems along the northeastern Atlantic coast, including the Connecticut, Hudson, Raritan, Delaware, Chesapeake, and Rappahannock, with many of their lesser tributaries. The associations of plants and animals described in any one zone of the river are specific and typical, but are only briefly representative of many other lives that might be present. Terminology has been kept as general as possible, at times even obscuring the identity of an organism or geological event, yet always referring to specifics upon which the entire narrative is based. Omissions far outnumber the little that has been arbitrarily included. So little of the whole is considered here, a reader may join me in feeling sympathy for Isaac Newton's recognition of the impossibility of grasping more than a trifle when he wrote, "I feel like a child who while playing by the seashore has found a few bright colored shells and a few pebbles while the whole vast ocean of truth stretches out almost untouched and unruffled before my eager fingers."

Ideas included here come from so many sources no proper recognition can be given in a book of this sort. I hope I have correctly interpreted the original work of others; if not, the responsibility is mine alone. The list is far from complete, yet I must acknowledge the early inspiration provided by the late Drs. Ulrich Dahlgren, Megumi Eri, and Thurlow C. Nelson; the frequent exchanges with good friends and colleagues, Drs. N. J. Berrill, Melbourne R. Carriker, L. Eugene Cronin, Franklin C. Daiber, Gairdner B. Moment, Carl N. Shuster, Jr., and Robert W. Stegner; finally the clear insight of my former students or assistants, among whom Drs. Joseph P. Bran-

ham, John C. Ferguson, Thomas R. Hopkins, Charles M.
Lent, James P. Thomas, and Messrs. Dexter Chapin and
Douglas C. Wilson have been foremost. Other authors, spe-
cialists, students, and my family have contributed constantly and often unknowingly, although they are sure to
find familiar thoughts scattered through these pages. Perhaps they will forgive the appropriating of their ideas in
view of the larger purpose of this book, for I suspect all of
them share the conviction that this is a glorious planet on
which to live and one worthy of preservation and restora-
tion. We may at times be blind to its beauty and to its
needs, but those astronauts who have looked back to our
"good earth" know well it is our only possible dwelling
place in endless, inhospitable space. Its brilliant blue and
white, water in all its forms, beckons home even those most
adventuresome of men.

W. H. A.

Manoa Valley
Oahu
1970

CONTENTS

xiv   *Contents*

IX

THE RIVER SET FREE 195

X

THE FULL CIRCLE 221

XI

THE INTRUDERS 245

Epilogue 264

THE
INFINITE
RIVER

Clouds, ever drifting in air,
Rise, O dewy anatomies, shine to the world in splendour.
Upward from thundering Ocean who fathered us
rise, make way to the forested pinnacles.
There let us gaze upon
summits aërial opening under us;
Earth, most holy, and fruits of our watering;
rivers melodious, rich in divinity;
seas, deep-throated, of echo reverberant.
Rise, for his Eye, many-splendoured, unwearying,
burns in the front of Heaven.
Shake as a cloak from our heavenly essences
vapor and rain, and at Earth in our purity
with far-seeing eye let us wonder.

ARISTOPHANES

I

FROM WHENCE
THE RIVERS COME

*All the rivers run into the sea; yet
the sea is not full; unto the place
from whence the rivers come,
thither they return again.*

ECCLESIASTES 1 : 7

Far out in the eastern sea, twelve hundred miles north
of the equator, a layer of gently moving air stilled the
ocean. Great blunted swells rose and fell in dreamlike
rhythm, attesting their origin in a distant storm. Single
translucent waves, larger than most, foamed momentarily
along erected crests, then lost their whitened tracery as
spume collapsed and disappeared in the shadowed slope of a
following trough. The sun hung alone in the clear afternoon sky, warming and illuminating the ocean.

Below the surface several finback whales swam, herding
before them an unseen prey. Slow, measured strokes of
their wide flukes drove the immense animals forward with
an ease that seemed disproportionate to the volume of
water which had to be thrust aside. Playing across their
dark bodies were bright, shifting patterns of sunlight that
filtered through the surface wavelets. The whales held open
their deeply cleft jaws, revealing cavernous mouths into
which quantities of water containing small marine creatures

were drawn. Beneath their mouths capacious pouches swelled with incoming water and food, until one after the other the whales closed their ponderous jaws, leaving exposed only a coarse fringe of short yellowish fibers, the baleen through which water was forced outward by pressure from their massive tongues. A multitude of planktonic animals strained from the water was retained and swallowed. The finbacks were grazing with little effort; their food, small in the size of its individual constituents, never lessened in its volume spread throughout the upper sea.

The blue expanse in front of the whales was transformed suddenly into a sparkling array of silver meteors which approached the surface in rapid ascent until each bright dart resolved itself into a small shining fish. Thousands of round-herring leaped free of the water, their glistening bodies held for a moment in the still air, twisting and shimmering, before they spattered back into the ocean, fracturing its calm. They did not leap again, but escaped the migrating whales which continued their steady course toward Arctic seas.

The round-herring quieted near the surface and schooled slowly, now intent upon their own problems of feeding, turned in an instant from potential victims to predators. With small mouths working quickly, they collected quantities of shrimplike crustaceans which formerly had appeared only as dancing glassy ornaments in the shifting rays that filtered through the surface. The pattern of escape reasserted itself in a behavioral echo as the tiny shrimp leaped clear of the water and, falling back, dimpled its surface, where they were caught in thousands of gulping mouths. For a while the attention of the herring was directed at the crustaceans, but as the two groups began to separate, the fishes continued feeding upon even smaller animals and plants which were either immobile or swam so feebly that capture was certain. Some of the minute beings also shone momentarily as they interrupted bright beams of light during their diminutive travels. A few of the quiet ones possessed natural buoyancy with spiny flotation devices or oil globules within their bodies. The safety of their kind lay in incalculable numbers—despite the millions being eaten by the round-herring and other filter feeders, there would always be enough to perpetuate their kind. Some of the small creatures were crustaceans, others were slender worms, and even very young fishes, big-eyed and transparent. Microscopic plants, housed in geometric containers of glasslike silica, floated in the mass of life that stretched beneath the swell of the sea.

All these living things existed in surface waters that had been heated for weeks by the summer sun. The warmth was a fluid blanket covering large areas of an ocean that seemed, without terrestrial reference points, forever stilled. Nevertheless, the creatures and the warm water in which they lived were being carried along an oceanic highway toward the shores of a continent to the northwest, transported by a vast current which obeyed forces created during the rotation of the earth and by the heat of the sun. Above the sea, warm air drifted with slight pressure toward the southwest, rising slowly with a languid turbulence.

Here at this scene, on a scale infinitely smaller than the most diminutive creature, and at this instant, a mighty river was being conceived on the quiet ocean surface through the union of a shaft of midsummer sunlight and particles of a receptive sea.

The story of a river is one of life: the existence of trout and cloud, of seaweed, torrent, dragonfly, raindrop, and starfish—the warp and woof of three billion years.

One thread of the continuum in time and space had begun long ago in the hidden core of the contracting sun; there, during a self-sustaining transmutation of elementary particles, a fusion of hydrogen to helium, powerful forces commenced their tortuous and often delayed journey outward to the face of that incandescent star.

On this day, after millennia of prolonged and erratic wandering within the sun, radiant energy and a surge of subatomic particles emerged from giant convection currents into the star's outermost flaming turbulence, were liberated and flung at the earth, ninety-three million miles away. The electromagnetic energy radiated in many wave lengths, from short gamma to long radio waves, with those of highest frequency partly reduced during their passage through the sun, and further retarded by a gaseous and luminous halo, the corona, which enveloped the sun. Although only an infinitesimal amount of the remaining energy bathed the small world which intercepted its path so far away, it created the river, gave light and warmth, and ran the life-engines that grew facing the sky.

Eight minutes after leaving the sun, blasts of radiant energy and corpuscular streams of atomic fragments began passing through strongly radioactive belts of electrically charged particles encircling the earth, the two major belts held thousands of miles out in the earth's magnetic field. Continuing down, the streams increasingly struck the first particulate traces of an atmosphere so dispersed its motes were ten billion times rarer than air molecules pressing directly upon the surface of the planet. These circulating atoms—first hydrogen, then helium and oxygen—were widely enough separated to travel hundreds of miles around the earth before colliding with one another. When they finally did, depending upon the violence of such a collision, some atoms sprayed out into space, others continued to be held in orbits around their parent world, and some were pulled downward where they recombined into molecules. Because of their electrified nature, even smaller atomic fragments resulting from these chance meetings were also firmly captured by the earth's magnetism.

Among the particles streaming from the sun were bare protons of hydrogen which joined with other particles in the upper atmosphere to produce more hydrogen, eventually creating at lower levels as much as one and a half tons of new atmospheric water a day. In this way, the sun contributed directly to the supply of earthly water and eventually to the river itself.

The huge vaporous shell embracing the world was held close against it, each minute gaseous particle pulled toward the compact earth by the attraction of one substance for another, until a majority of gas molecules nestled in profusion upon the surface of the globe. This was the dense lower atmosphere in which the river began, an enormous, churning mass of gases and of suspended liquids and solids, all set in motion by the sun's energy. With its clouds and air-borne water at the lower levels, the atmosphere made rain possible; its oxygen and carbon dioxide permitted the existence of life which breathed in one gas or the other; its density conveyed sound and provided opportunities for the development of flight by animals and passive seeds. Thirty miles above the earth, the atmosphere already was dense enough to consume meteors in a fiery plunge, protecting the surface below. Thick though it was, the airy envelope allowed light and warmth to penetrate, yet reflected excess amounts of each, maintaining underneath a climate which was favorable to life.

Despite their nebulous character, the outermost shells of air, the exosphere and ionosphere, shielded the world beneath from a tremendous electromagnetic fusillade which battered constantly at the earth's frontier. The activity of the vague, heated regions of the upper atmosphere, shaped by the force of the solar wind, created stormy fluctuations of electrical forces as they responded vigorously to the unending bombardment from the sun and outer space.

Perhaps the most important feature of the ionosphere was that of a guardian of life below, for it absorbed harmful radiations from the sun and resisted cosmic rays from unknown regions far beyond the solar system. It was here that lethal ultraviolet radiation from the nearby sun was first absorbed, its energy spent in breaking apart molecules into atoms, and atoms into lesser particles possessing electrical properties. The more powerful cosmic rays, possibly coming from the explosion of a distant star, struck the atmosphere every second with tremendous force. As a cosmic ray collided with atoms of air, energetic particles showered downward in a cascade that dispersed and did little harm to life, leaving high above a region of ionized, electrified fragments of atoms and molecules. Only one cosmic ray in every hundred thousand reached the earth's surface at anything near its original strength, yet over a period of three and a half billion years these rays, in addition to ultraviolet from the sun, high temperatures, and other factors, had caused slow, random genetic changes in the life begun so simply and precariously in shallow seas after molecules became able to make more of their own kind.

The filtered and reduced radiation from the sun descended into the more dense, weatherless stratosphere, a region possessing jet winds of far great velocity than those of the lower atmosphere. There ultraviolet rays created a three-atomed gas, ozone, which absorbed enough of this radiation to prevent excessive damage to life on the surface.

Despite absorption, reflection, conversion, and other forms of opposition, the sun's energy continued to approach the earth, speeding through the stratosphere into lower levels where the swirling, turbulent air of the troposphere, weather's own seething caldron, held nearly all atmospheric water. The concentration of water vapor differed over the face of the globe, but together with carbon dioxide gas it absorbed the long waves of heat rising from the earth's surface below and radiated it back downward again, in large measure balancing the planet's heat loss and maintaining its climate.

Because of collisions with air particles, the now-slowed rays from the sun were forced from their parallel paths and became erratic and confused. Over a third were reflected immediately back into space; some were scattered by bouncing irregularly from particle to particle; and a few decelerated even more and were converted into the long waves of heat. The remaining radiation, mostly visible light, pierced the lower atmosphere to descend to the ocean far below. A clear summer sky allowed the rays to continue with little further interruption.

Light—radiant energy discernible to living eyes—was now scattered and diffused, securely trapped in the blanket of air, creating a vast luminosity that gently lighted the cradled world beneath. In the lower air the cumulative effect of minute particles of water, salt, minerals, spores, and dust was to create a pale blue sky. Later in the day, with the sun low and its oblique rays struggling to penetrate the atmosphere, the trapping of light would be so complete that only the longer wave lengths of red and yellow would succeed in ending the daylight hours in sunset glory.

This was the energy that had reached the blue and white earth long after originating deep within the sun; that had raced uninhibited through ninety-three million miles of space in eight minutes; that had been slowed and deflected as it passed through a blanketing atmosphere. This was the energy that had taken many forms, yet only entered the river's story as it underwent its final transformation from radiation to an invisible, working force, capable of evoking a change in the excited molecules of the earth's surface.

From high above, the translucency of the still water seemed to have no limit. The sun's rays appeared to penetrate to the depths of the oceanic abyss: such is the illusion that clear water presents while it diffuses and absorbs light.

The color of the open sea, like that of the sky above, was a pervading blue, also a result of the scattering of light rays. Toward distant shores to the west and north where water was less clear, light rays shifted to those of longer wave lengths, becoming green and yellow and—ultimately—a turbid brown as large quantities of silt were carried from the land.

Light striking water can go in only two directions: downward, or back into the sky. Light rays actually entering water tend to turn down, sinking straight into the ocean. With the sun high above, as it was this day at noon, light stabbed through the upper sea, a few rays reaching great depth. Later in the afternoon, more light would be reflected off the surface, until by late in the day most light illuminating the sea would not penetrate it, but turn outward.

The retardation of light in sea water was more rapid and effective than in the less dense atmosphere. Within the first few feet almost two-thirds of infiltrating light was slowed, converted to heat, and dispersed by its impact against suspended matter and the water molecules themselves. This heat did not persist, but decreased so rapidly that a condition of perpetual cold existed at medium depths. Light, too, was reduced quickly and eventually extinguished. The few infrared and ultraviolet rays which remained after passing through the atmosphere vanished almost immediately upon entering the sea, and those of visible red and yellow were greatly diminished. Below fifty feet there was no red light; yellow disappeared at three hundred feet. At eight hundred feet much of the blue-green vanished, leaving only a midnight blue. By seventeen hundred feet the water was devoid of light to most eyes, although faint radiation continued for a thousand feet more. The remaining depths of the ocean were utterly dark.

The sun's energy pouring into the water this very instant exceeded four thousand horsepower for every sunlit acre. The upper ocean, subjected to such vast quantities of heat for months, had grown slowly warmer, until now the surface stood nearly five degrees above the temperature six months earlier, when the sun had begun its climb to higher latitudes. The ocean was an immense reservoir of heat energy, transporting and circulating its warmth over a wide expanse.

Oceans do not retain all their heat. Only in one way can the earth as a whole rid itself of the gathering energy: by allowing it to radiate into surrounding space; but it would do this entirely too freely were it not for the insulating effect of the atmosphere.

On this bright and almost cloudless day, a tenth of the earth's heat successfully penetrated and escaped the gaseous envelope which swaddled and cultured the warm world beneath. The rest was retained in the air, where it assisted in its own circulation through the creation of air currents, the winds. Energy levels on earth had assumed a stabilized nature ages before; changes occurred only with the cycle of seasons and the slow development of climatic eras. The minor fluctuations of temperature at the ocean's surface were but a part of a timeless constancy.

The warmed surface water composed a buoyant layer separated by a temperature barrier from the dense bulk of the colder ocean beneath. The barrier, although invisible and changeable, effectively blocked transmission of minor currents in the water, and even affected to some degree the vertical migration of certain marine creatures. Closer to the continental shore, these same surface waters would be displaced by colder water from the depths through an extensive upwelling, an overturning that would mix waters which formerly had remained distinct. Mixing did not occur here, however, for a massive current, over fifty miles wide at this point, stretched across the scene, flowing northwest toward the still-distant continent.

The current formed a spacious highway for the transport of energy, materials, and living things. The warm, homogeneous mass of water composing the great current would later affect the climate of lands past which it would flow:

to some it would bring rain, and for others it would temper
the rigors of northern latitudes. Far to the north it would
lose much of its warmth, but eventually would flow south,
deflected by the shores of a distant eastern continental wall,
until the forces of the rotating earth would again propel it
westward across the sunlit ocean close to the equator, com-
pleting its circuit in a year's time.

The path of the current was not always regular: it would
fade and reappear, and often double back upon itself, forming gigantic loops which might break from the main current and be lost, assimilated into the ocean's bulk. The
entire current encircled a quiet and impoverished sea,
where life was sparse.

The streaming current was not deep, but its great width
and steady flow transported over one hundred cubic miles
of water every hour through the scene. Beneath it there
flowed a countercurrent, separated not only by its direction
but by a temperature different from both the warm water
above and the colder mass beneath. Very far down, three
thousand feet and more beneath the surface, the water was
more nearly at rest and had little communication with the
upper strata. Here was a perpetual coldness.

Despite the heat that radiated back into the atmosphere
and the frigidity of the deep ocean, most heat energy
entering the sea remained, and some of it would soon be
used in the creation of the river. The power, the very life
essence of the river, was building and beginning to stir.

An ocean is not water alone. In the young world when
this fluid condensed as the earth formed and issued from
volcanic origins, collecting in rocky basins, the early seas
might have contained fewer minerals, but probably it was
salty even then. Soft minerals in the basins would have
dissolved at once in even the first traces of moisture, for
water is close to being a universal solvent. Over the ages, as
water ran across the face of the land, a multitude of additional mineral compounds and electrically charged particles

were carried to their oceanic destination. Some substances leached upward through the sediments of the sea bottom, while others were washed out of the shores by wave action wearing away the land. Over three percent of the ocean's mass is now composed of various combinations of forty-nine elemental substances.

Should all these materials in solution be removed, water itself would remain a unique compound. It is extremely cohesive to itself, and adheres closely to objects and dissolved particles which happen to be present. So powerfully does water cleave unto its own substance that it has a strength equal to that of some forms of metal. With the combined effects of cohesion and adhesion among the particles in solution and the fluid itself, ocean water tends to resist any change in form and movement unless driven to do so by powerful forces of wind and current, with the result that working energy is imparted to the water. Waves crashing upon a distant shore where they shear cliffs into rocks, and crumble rocks into sand, bear witness to the power they receive from the atmosphere.

In the region where the river was conceived, the viscous quality of the water greatly enhanced the support of many small living things that had to remain close to the surface or perish for want of light and warmth. To these minute beings—the larval fish, worms, plants, and crustaceans that had been pursued by the round-herring—the sea had real substance. It was a dense medium in which to feed, flee, and reproduce.

Far smaller than these minute creatures, tiny motes of the river joined the corpuscles from the sun in being the elemental trifles that compose all material existence. They were the atoms and their component lesser particles. In locked and united clusters they formed molecules that moved spontaneously, not with life, but with the energy contained within their unifying bonds. Large molecular aggregations of various substances, suspended in the water, suffered a continuous and irregular bombardment from the smaller water molecules, until there was a vast and endless twinkling in the sea as each diminutive particle was jostled into reflecting a bit of sunlight. Radiant energy increased the activity: when long heat waves struck individual molecules, especially those of water, the atoms of which they were composed grew intensely excited. Each hydrogen atom was moving faster than a mile a second, yet was surrounded by equally active particles that interrupted one another's progress. The furious pace of broken and re-formed hydrogen bonds created a violent and chaotic condition which would slacken only during the night and in the winter to follow.

Not all water molecules moved at precisely the same speed. Some were sluggish, while others were unusually energetic. In the frenzied movements of the closely packed minuscule bits of matter, repulsion and attraction were almost equal forces in the sea, but near the surface the uppermost molecules were held tightly by bonding forces only from beneath, where the bulk of water lay. Attraction of water for itself resulted in an elastic film which stretched over the ocean, an almost impenetrable ceiling to the very small creatures that lived underneath. Gas molecules passed easily through the film, seeking a state of equilibrium, a condition never achieved because these same gases were constantly being circulated and used by the sea and its inhabitants, then released back to the atmosphere in altered form.

Molecules of the warmed surface performed their ceaseless jiggling dance. Some of those in the undulating film, responding to violent collision, bounded into the air above. Many of the less active molecules were pulled back into the sea at once, but a few passed beyond the grasp of the ocean, were catapulted still farther into the air, and were held. In this way, water rose steadily, carried aloft by diffusion, rising air, and winds, its rate determined by subtle changes in temperature and pressure.

Without heat, water molecules cannot move; without energy, water cannot be moved. Wherever there are movements of large volumes of water, there is a prodigious transfer of energy. During the evaporation of water from the sea and its ascent into the atmosphere, there was an immense, wholesale removal of heat into the sky. At this midday hour almost every vestige of surplus heat in the surface water was being used in evaporation. The region of the sea where the river began was making a substantial contribution to the millions of tons of water which were rising from the world's oceans that very instant.

Immediately adjacent to the ocean there rested a thin layer of air heavily laden with water molecules only just departed from the sea. This layer assumed the same temperature as that of the water, acting as a zone of transition between the densely packed molecules of the sea and those which were fully air-borne. The thick vapor held almost immobile to the surface by friction not only brought with it heat from the ocean, but acquired even more warmth of its own, some from short-wave sunlight, and some from long-wave heat radiation. This energy-rich water vapor, now held in the atmosphere, rose in small aerial eddies, its energy not to be released again until many days and many miles had passed.

The wind, gentle though it was, continued to blow toward the northwest and gradually was replaced by an air mass cooler than the sea, a condition which greatly enhanced evaporation.

With the most energetic water particles leaving the sea in quantity, there was a perceptible drop in water temperature at the surface, causing a heaviness. Furthermore, dissolved salts, left behind during the evaporation of water, also accounted for an increase in the weight of the water. The

cooler, more dense surface water then began to sink and
was replenished by warmer, less heavy water from beneath.

Hundreds of miles to the west, large rising air currents
had been developing over the distant continent, and their
effects were now apparent over the ocean. Cool air was
being pulled down from high altitudes and began to flow
ever more swiftly across the sea toward land to replace the
hot continental air which was billowing upward in violent
ascent. During its passage across the ocean, this air grew
humid and warm, developing characteristics of its own,
quite distinct from neighboring air masses.

The hastening breeze accelerated the process of evaporation. As soon as water molecules diffused into the air above,
they were hurried away, borne upon invisible eddies which
carried them high into the atmosphere where they were
widely dispersed. The farther aloft they were conveyed,
the more rapidly they flew to the west, suspended in the
stream of turbulent air escaping from the retarding friction
of the sea.

The vibrant energy of the sun had joined with the fertile
sea to produce one of the seeds of the mighty river.

Smooth swells began to ripple from the stirrings of the
new breeze. As the wind built, wavelets grew slowly into
crescent-shaped, scalelike waves which became higher and
more separated as the hour progressed. There was a rustling
at the surface, a murmuring that grew across the wide
expanse of ocean, punctuated occasionally by a slight bubbling splash as a wave crest momentarily foamed. The wind
had an unlimited sea to work upon and it pushed and
tugged at the resisting water until waves began to accept
the brisk activity of the sky-borne energy, and rolled and
broke, flinging racing white fingers down their leeward
slopes. Small areas of surface were shattered into tiny fierce
catspaw ripples which chased and circled, blended and
crashed headlong, driven by fitful gusts of the rising wind.

From close on, the waves appeared to parade in even file, but from high above, the sea was a sheet of crinkled blue, shadowed and dashed with white, an irregular pattern of wave elevations and depressions, a topography wild and disordered.

The waves grew in size as violent energy was transferred back from atmospheric wind to water. Each wave was an obstacle to the wind blowing fast against it; as the pressure increased on a wave's windward side, it grew taller and hurried on a little faster. But the water itself went nowhere, for a wave is but a form of vertical movement. Even this oscillatory coursing did not do the bidding of the wind by flowing precisely in its direction: the unconforming waves proceeded in their own manifold fashions.

As waves broke, droplets of salt water were wrested from them and flung into the air. Some fell back, their splashes dimpling the fretful surface. Others were suspended in gusts of wind which rushed up wave slopes and continued aloft. Turbulence in the air carried the drops higher into regions of dry air where they evaporated when their individual water molecules were wrenched from one another and dispersed. After the water was gone the salts which had been dissolved now were exposed. Here they remained suspended in great numbers, perhaps a thousand to a cubic inch, and were responsible in part for the scattering of blue rays throughout the sky. Much later and far away, these same salt crystals would become an important factor, an omnipresent catalyst, in the birth of the river.

High in the clear sky the invisible separated particles of salt and water tumbled about, driven by the windstream.

II

UPON THE WINGS
OF THE WIND

I am the daughter of Earth and Water,
And the nursling of the Sky;
I pass through the pores of the ocean and shores,
I change, but I cannot die.

PERCY BYSSHE SHELLEY

INTO the tenuous and far-reaching sky passed the particles so recently risen from the sea. Here they flew through an airy world which would hinder and delay their deposition upon the waiting earth below. Only through the sky does water reach a continent, and only as vapor can it pass beyond the shore which holds back the ocean.

The dense lower atmosphere containing the embryonic river pressed against the earth's surface with a force of five million billion tons, yet it was not still. Circulating and mixing in wind currents created by warmth drunk in by land and ocean, it responded to forces from beyond the earth. As the moon swung about its planet, ponderous swellings in the oceans followed, lagging behind its piping call of gravity. Tides in the atmospheric sea were present too, but were concealed in the lower regions of the airy shell where they barely acknowledged in slow pulsations the activity of both sun and moon. Only far out in the ionosphere were moon tides pronounced; there they rose and fell over a mile in height.

In the area where water molecules had been rising, winds flowed out of the east, quickening after a period of relative quiescence, reaching their greatest velocity a mile or two above the ocean. Day after day they streamed, tumbled, and blew, pushing on in hurried and urgent progression, a constant and reliable engine of motion.

In the equatorial regions a thousand miles to the south, the sun's heat had set the atmosphere in motion, while the spinning of the earth had fashioned it into a complex system of whirling eddies. Two enormous, earth-girdling masses of rotating winds, the prevailing easterlies and westerlies, were not only the agents of transport for the river, but its very womb. Within these winds the seeds of the river would be nurtured and grown until time for their emergence as drops of water.

Within its system of planets and sun, the earth is blessed by lying in a zone which allows water to exist primarily as a fluid and only to a lesser extent as a vapor or solid. In fact, water is one of the few natural substances that exists as a liquid on the surface of the earth; another is mercury, a highly poisonous metal. Life is inextricably linked to this condition, for living organisms may be found only in water or in moist air close to the floor of the atmospheric envelope.

Water is a sturdy molecule, requiring a great deal of energy to break it apart, although its elements of hydrogen and oxygen come together quite easily. Because it is a supreme solvent, it seldom is pure but carries about a vast array of dissolved substances, many of which are essential to life. Most of these it obtains in its role as a powerful agent in the shaping of the earth, descending upon the land and running toward the seas, leaching minerals from the soil as it goes. Water's physical properties are no less important to life than those of a chemical nature: it is nearly

incompressible, requires much heat to vaporize, and is slow
to give up the vast quantities of heat it absorbs. The fric-
tionless sliding of one body of water upon another makes
internal currents possible and the strength of its surface
film, important to so many plants and animals, is exceeded
only by that of mercury.

Life originated in water and every living cell, from the
beginning to the end of biological time on this planet,
requires water as an essential component in its chemical
machinery. The complex bodies of larger plants and animals function only by means of watery solutions and hydraulic transport systems.

Toward the outer limits of the narrow zone of life set by
water, high-velocity jets of wind shape clouds into their
ethereal patterns. Closer to the surface, winds might attain
speeds of thousands of miles an hour were it not for the
retarding effect of friction. Lower clouds simply follow the
unfelt speeding of the earth itself. The circulation of the
atmosphere allows life to exist under equable conditions, for
without the moderating effect of the winds, the equator
would be unbearably hot and the northern latitudes intol-
erably cold.

Wind patterns are shaped not only by heating and the
rotation of the earth, but by continental land masses. General wind directions usually follow coastlines, with fluctuations on a smaller scale each day and night. During summer daylight hours, air ascends over the warmed land
drawing in winds from the sea; at night the constant ocean
is relatively warmer, and air is pulled out to sea.

The immense amount of water vapor borne by the winds
was being replenished as rapidly as it was carried away.
Each hour tens of thousands of gallons ascended into the
atmosphere from every square mile of ocean, until within a
year's time almost fifty thousand cubic miles of water
would have risen from the world's seas. Much of the water

would fall back into the ocean as rain, but nearly half would be carried along, as was happening here, hundreds and thousands of miles before being showered down.

As the evaporated water molecules rose into the streaming air they became widely separated into a lessened pressure, until at a height of two miles they raced in invisible isolation toward the northwest. This vapor, originating in the central latitudes of the ocean, was to become the one most important source of rain over the entire eastern half of the great continent which lay ahead.

Tumbling about in the wind, the newly risen molecules passed and collided with a multitude of older, more seasoned particles—bits of smoke, dust, salt crystals, and an occasional mote of life. The smoke particles came from long-extinguished conflagrations on the earth's surface, the dust from the land, volcanoes, and from meteors consumed in the air, and the crystals, most numerous of all, were those which had come from the ocean.

As the irregular coastline of the continent was approached, the once invisible crystals began gathering into a faint and luminous haze, glittering in the sunlight. Each crystal had enlarged slightly, swelling with the addition of water molecules so abundant in the maritime air. As this absorption continued, there was an increased misting and whitening of the blue sky.

Minute spores and cysts were the first forms of life encountered by the embryonic river during the early hours of its existence, yet it was a chance meeting only, for the air was too turbulent and warm to allow a union of life particles and water.

Living things in the high atmosphere took a variety of forms, although all were small; so small in fact, even in still air the surrounding pressure would keep them suspended for days or years. The most abundant were geometric in shape, often with spines, blunt points, or stubby wings:

these were pollen grains, containing the lost sperm of earth-bound trees, encircling the world far from the passive egg held deep within a flower.

Other specks of quiescent life were bacterial spores, the resting bodies of microscopic organisms which live every-where on the earth's surface and in its waters. Within the protective walls of spores their minute lives awaited the passage of time until the vagaries of wind carried them down again to a favorable and receptive world. With them in the airstream were the larger spores of fungi, fewer here than over the land which was their place of origin. Many of the spores would never germinate, their lives terminated by too long a passage in the skies and by overexposure to the lethal rays of ultraviolet which had filtered through the outer shells of air.

At times larger and more irregular objects were carried past in the windstream: protozoan cysts, tiny encapsulated animals, blown from the dried, dusty bottom of a distant pond. Even in protective cases their lives were short, and should they not soon fall into moisture, each cyst would become an empty, infertile relic of an insignificant and transitory life.

The capricious wind caused its passenger list to be un-predictable. Some of the suspended objects had been in the wind-flow for weeks, others for years. To the southeast, a few islands lay in the region of the strong tropical updrafts which had set the great wind belts in motion. Minute aerial flotsam had been sucked from these islands and now, far to the northwest, bits of straw, volcanic dust, fragments of plant fiber, and dead midge-sized insects passed by in never-ending procession, a hurried review of all the trivia that comprise a world.

The ocean over which the wind-driven motes and parti-cles passed had a character quite different from the open sea. No longer was it a transparent blue but had grown green and opaque. More fishes were feeding close to the surface, and on the horizon bottle-nosed dolphins rolled slowly in and out of the waves. The waves themselves, responding to the effect of a shallower bottom as the continental shelf rose beneath them, grew taller and sharper. Although surface winds had abated, waves continued to chase up and down the increasing ground swell.

The wind passed with the hours in this open world, and the scene shifted slowly to the northeast. In places the green sea was shot through with long yellowish tendrils of discolored water which held traces of the land—sediment, a bit of wood, a fruit, and fragments of marsh grass.

The atmospheric haze which had formed high above the coastal waters was consolidating itself into wisps of clouds, not clearly defined yet, but building surely. Salt particles added to their first water molecules great numbers of others, until the salt no longer was a nucleus for condensation, but had itself dissolved within a bit of accumulated water, a microscopic droplet still so small that gravity had no attraction for it, and it was borne on the wind without effort or resistance.

As the air cooled slightly, the activity of its water molecules lessened. When they came into contact with surfaces, even those of infinitesimal particles, they did not rebound but clung fast by means of their hydrogen bonds. A developing droplet did not always require a crystal, spore, or bit of debris to condense upon; many droplets formed simply by the coalescence of water molecules. Within minutes distinct clouds were building, until rank upon rank of puffy, elongated forms blew in from the sea.

Deep in a cloud the first active life of the river appeared. As molecules joined and condensed into a droplet around a spore, the case broke open and a bacterium quivered into life. Shielded from the sun's lethal ultraviolet rays by water vapor in the cloud above, it emerged from its protective case, and with quick vibrating activity moved from one side of its glistening prison to the other. Long before the river itself was born, it had offered life to one of its smallest companions which for a brief while swam about in its hastening sky-borne vehicle.

The air flowing in from the ocean this summer afternoon brought with it characteristics of temperature and humidity acquired during long contact with the sea. For the time being, as maritime air passed over low-lying coastal plains, even more evaporation took place and water left salt marshes, streams, ponds, and the moist soil itself. The dark soil of the alluvial plains absorbed much of the sun's heat, warming the air above and causing it to sway in the dance of a mirage. When the earth was heated and wet, it often gave rise to fog and mist which carpeted the coastal plains at night.

The air mass from the sea was still all-powerful, but across the land other forces were marshaling: air masses from the dry, hot continent and from the distant, chilled polar lands. A collision course was set, a struggle for survival would take place, with the sea's legions falling to earth in defeat.

Also waiting inland were huge towering sentinels, bulwarks of rock against which the hastening air from the sea would press and grow confused in the currents which seethed about the mountains.

Before meeting these opposing forces, the maritime winds had to flow over a sun-drenched land which sent massive pillars of hot air rising to deflect and rob the streaming air of its watery cargo.

The expanding air, growing lighter, rose in updrafts at speeds of over a thousand feet a minute. Everywhere across the plains were great, invisible columns of air ascending in regular order and with equal velocity, a colonnade of vertical winds.

At five thousand feet, where the air columns had grown cool, condensed vapor sketched a gauzy, transparent ceiling. It was vague at first, but as droplets collected and grew larger, it resolved itself into a distinct stratum of elongated cloud-wisps which at first resembled the clouds formed earlier at sea. As the air rushed upward even more rapidly, the clouds began to build upon their flat bases. Mass upon mass, billow upon billow they rose, forming, parting, joining, swelling, finally towering in blinding, snowy majesty above the now shade-spotted plains. Each huge cumulus emerged as the structured crown of an invisible current sent aloft by rushing winds.

Within half an hour the clouds towered thousands of feet above their flat bases, each cloud containing over a thousand tons of water suspended in countless minute droplets. The droplets, less than a thousandth of an inch in diameter, now ceased coalescing and, buoyed by updrafts, drifted inland with the sea breeze.

The smaller sea clouds mingled and became lost within the enormous cumuli, and what few bits of debris and life they contained now were joined by quantities of dust and organic particles sucked from the land. Many of these recent arrivals served as nuclei for further condensation of water droplets, and the number of tiny lives liberated into cloud-water increased. Bacteria, one-celled animals, and the cells of primitive plants emerged, some to die at once, others to survive for a time. From the waters of tidal creeks and inland waterways came infinitesimal living things which had been pulled aloft in spray droplets without having had a chance to cover themselves with protective cases. Most died quickly, but a few lived on. Now there were even larger forms of life in the clouds, those which had little to do with water. All manner of flying insects, having ventured too far above the ground, had been caught up in the violent updrafts and thrown into the clouds where they were whirled about, no longer capable of controlled flight. Wind-driven seeds of dandelion, milkweed, and other plants rose and danced in the turbulent air.

Beneath the clouds, in the clear buoyant streaming of

vertical winds, hawks soared, carried upward boisterously almost to the flat shadowed ceiling, then gliding, careening, plummeting down along paths of flight to their world beneath. Few birds penetrated the cloud mass, but left the huge white mountains and their bottomless valleys desolate of creatures which could explore at will the ever-changing contours.

Rank and file, the flat-bottomed cumulus formation sailed inland from the sea, invading the entire coastal plain for miles to the north and south. On the earth beneath, great silent shadows came and went as the sun was obscured and then released. The sun itself seemed to sail through the clouds, changing from a luminous white disk to blue to deep orange before disappearing totally behind a massive cloud bank. Most of the cumuli were brilliantly lit, their contours revealed by subtle shadings. When a cloud drew over the sun, it became dark and dirty, with only its margins etched against the sky by a golden light. Rays broke through at times and fanned across the sky.

The clean outer brilliance of a cumulus vanished inside the cloud where it was as cold and damp as fog, murky and gray, with ragged wisps of vapor streaming in the swirling air. This was the dark home of a growing power, sky-held energy of the distant sun which later would cause the cumuli to build into thunderheads stretching miles upward into the atmosphere, their energy ultimately released as heat and plunging rain and lightning.

These enormous clouds, the writing of wind upon the sky, dramatized the currents and turbulence of the lower atmosphere, foretelling events that were to happen over the slopes of foothills and mountains guarding the innermost continent.

III

NATIVITY

For he maketh small the drops of water: they pour down rain according to the vapour thereof.

JOB 36:27

B Y late afternoon banks of cumuli stretched in unbroken array across the plains, covering the foothills, their vanguards pressing against the first mountain slopes. Their progress partially checked, clouds and the whole layered air mass slowly began to rise. The ascent created a cooling effect that turned many droplets into ice crystals, which in turn robbed water molecules from adjacent droplets.

Cloud droplets and crystals, once infinitesimal and unaffected by gravity, now grew larger and heavier and fell whenever there was a momentary relief from the updraft currents, only to be caught and blown aloft again as the winds recovered. The droplets especially were growing steadily and the time was imminent when even strong winds could not prevent their return to earth.

The vertical winds lost their regularity and permanency, growing turbulent and disordered. At the bidding of the confused air currents rising over the foothills, clouds tore apart, re-formed, and piled upon one another, losing their individuality. Lost wisps of cloud swirled about the branches of trees on the shadowed mountain slopes beneath.

The sun still illuminated cloud tops, but it was sinking rapidly beyond the horizon, and the mountains and their valleys grew dark without the customary summer twilight allowed by a clear evening sky. As night drew upon the hills, the first rain came, hesitantly at first, then in drenching sheets which heralded the arrival of a great event. Trees tossed and strained, but soon the early downpour subsided, and rain fell throughout the remainder of the night, softly, without violence, in a gentle fashion, yet one that held enormous and unremitting power.

This was nativity, the birth of a river conceived in the southern sunlit ocean and nurtured in the womb of the sky. There was no fanfare for the birth of the giant, no roar or gusts of sleeting drops, but over the entire expanse of the hills there was a steady descent of drop upon drop. The easterly winds which had brought moisture-laden air inland still blew far above, releasing their water in droplets that descended in diminishing parabolas. The entire mass of falling water had a stilling effect upon the air, until in the lower regions rain fell straight toward its destination upon the mountain slopes.

Everywhere there was a whispering, as water, created of the earth, returned once more. Hour upon hour the rain descended, the abrupt arrival of each drop recorded when it struck ground or leaf, compressing the air in a sharp, fluid report. Small sounds radiated upward and outward, blending and losing all individuality, until to ears that heard, the forests and lower fields rustled with a pervading, welcoming murmur.

A raindrop, a million times the size of a cloud droplet, would have been a perfect, glistening sphere if surrounded by equal forces. But this could not be, for winds and changing pressures buffeted it so vigorously that its restraining surface film could hardly hold the water it contained, much less maintain the shape of a spheroid. As it began to fall, opposing pressures underneath cushioned its descent, causing the bottom to flatten out and even to become concave, while the remainder billowed out above in a trembling, ballooning mass. When the drop flattened out, it side-slipped, altered course, then again dropped straight down. Set in motion, a drop responded to both internal and external stresses until it oscillated and quivered like a living thing, alternating between a globule, a fat parachute, and a flattened disk.

A single drop, indistinguishable from its neighbors, collided and coalesced with other drops as it fell. Although it was repeatedly borne aloft again by the wind, its rate of descent, governed by its growing size, kept accelerating. When it struck a drop of similar size, both shattered into a spray of smaller droplets which began the falling and rising process over again. But the number of heavy drops increased, and soon many of them were clear of the cloud and fell directly to the earth.

The rain created a luminous halo around the distant hills, but overhead there was only a scudding darkness. Haze in the treetops muted the color and outline of their foliage, and trees farther away were only faint outlines and at times totally obscured.

Few drops struck the earth without previous interruption, but were intercepted by trees, shrubs, and low-lying grasses. When drops broke against leaves and splashed off, they multiplied into innumerable little droplets, causing the entire lower canopy of woodland plants to quiver as rain cascaded from above. Every seedling, every fern of the forest floor trembled in a rooted dance.

Leafy trees shed water quickly, but the evergreens, glistening with silver drops, held much of what they received between tightly spaced needles. Hemlocks displayed glassy spheres at the end of each of their tens of thousands of blunt needles. Heavy drops collected along fir branches, but left at drip points halfway along, falling to branches below. Streams of water clung to the underside of maple branches in crystalline ridges. Every twig, every stem and branch in the forest was wet with a thin layer of moisture, and the entire film flowed steadily downward along the trunks until it entered the soil where it diffused outward through the loose, absorbent forest floor. The leaf mold covering the soil became a rich, dark-brown carpet thick with captured water. Rocks jutting above the ferns turned a deep gray, and the lichens encrusting their sides no longer were dusty and dried, but grew green as the algae within drank in moisture. At the base of the rocks, the green of mosses deepened and grew richer, and the little plants straightened with water taken in through their delicate rootlike fibers. A cluster of reddish mushrooms glistened and swelled with succulence.

Whenever a large drop struck saturated soil, its compression created a tiny coronet of two dozen or more little spheres which shot into the air, then curved and fell spattering back, describing a dimpled circle around the original impact.

When each raindrop was finally stilled, its covering surface film immediately spread over soil particles and bits of vegetation, until every solid surface had stretched over it an unbroken layer of water. Underneath, the water flowed quickly between grains of sand and earth and spread through the countless galleries and internal spaces of the soil. Often it followed the hidden tracery of roots, filled the burrows of earthworms, and flooded the dwellings of ants, grubs, moles, and snakes.

Hour after hour throughout the night the forested hills drank in the rain. After the water had infiltrated the loose organic soil at the surface, it sank below the last humus layers and encountered sand and gravel. All particles in the soil, organic or not, accepted a film of water which clung fast due to molecular forces operating between the particles and the water. As the film increased in thickness, spaces were filled and air displaced, although air bubbles might be trapped beneath an impervious bit of clay or rock. The soil had been loose with air passageways which had permeated it, but now was becoming heavy and dense with accumulating water. Atmospheric gases, once plentiful in the soil, were diminishing. Some small subterranean lives tolerated the water, some welcomed it, but others drowned and suffocated from the lack of oxygen. A few crept into the rare air bubbles that remained, and were safe.

Not long after midnight the falling rain began to exceed the capacity of the soil to absorb it. Water accumulated on the surface as the overflowing ground refused it, then began to slip into small depressions where it collected in shallow pools. Rivulets carried masses of brown fir needles down their turbulent channels. The needles formed miniature logjams, creating additional pools until the dams gave way with a rush. As dams broke and pools spilled over, the water joined other streams until, when the level of encircling retaining elevations was reached, sheets of water slipped like quicksilver down one gentle incline after the next.

When raindrops struck a pool's surface, small erect spouts jumped instantly from the point of impact. At the top of each spout, a tiny droplet broke free, shot higher into the air, and fell finally into the bombarded puddle of which it became a part. Occasionally when a drop struck water, a thin ring of muddy water arose, joined at the center, and trapped air in a large bubble which floated briefly until it was punctured by another raindrop.

The rain-soaked soil was of complex and varied nature. At the very surface was a layer of decaying organic matter, mostly vegetable in origin, but with some animal remains scattered about in the form of insect bodies, bits of bone, feathers, and other trivia. Beneath this, a layer of bacterial fermentation gradually shaded into rich organic humus which marked the lower limits of most roots and burrowing animals. Still farther down, the earth was composed of rock fragments, sand, and clay. The process of breakdown of minerals to smaller particles was an exceedingly slow one that had been occurring since the uplifting of the mountain slopes, two hundred million years before.

Water deep within the soil moved slowly. It sank down, percolating through the loose earth, where it touched upon a multitude of tiny lives which depended upon its presence. As water dissolved soil minerals, new conditions of acidity, alkalinity, and electrical charge were established, conditions affecting most of the small earth-dwellers. Mechanisms in the cells of roots began their chemical pumps, not only bringing in water but using it as a medium for the transport of dissolved minerals across their cell membranes. Fungi and bacteria emerged from spores minutes after being surrounded by the seeping water.

Bacterial populations in the soil were changing. Those requiring oxygen were giving way to others which became active only when that element was absent. This cessation of activity was only temporary, and later when the water would seep away, the oxygen-users would begin their life processes again. Bacteria in the soil had many functions: some worked upon mineral compounds of sulfur, nitrogen, and iron; some broke down organic compounds in the process of decay; and a few affected the growth of plants by producing antibiotics. In turn the bacteria were preyed upon by small soil animals, and attacked by viruslike bacteriophages, tiny motes which could destroy them completely by diverting a bacterium's own life processes. When conditions of moisture, temperature, and food supply became unfavorable, many soil bacteria could form resistant spores. At this moment in the wet earth, some bacteria were emerging from spores, while others were forming them.

Although many bacteria moved freely, their diminutive size prevented them from making headway against the slow currents as water flowed through the soil. By the billions they were carried from the surface to deeper strata, then horizontally as the ground water sought emergence at a lower level. No soil particle or interstitial space was without its own large and varied population of bacteria.

Throughout the soil, formerly inactive strands of fungus mycelium now grew vigorously, forming new branches from each of its delicate, gossamer threads several times in the space of an hour. Fragile as the strands of mycelium were, they penetrated the soil in all directions, most abundantly in the rich humus, but still significantly in poorer soils several feet below the surface. The primary function of mycelium was to absorb dissolved foodstuffs directly from the soil, but some of these fungi were engaged in trapping microscopic one-celled animals, while others caught active roundworms wriggling through the soil water from one crevice to the next. The latter fungi had grown small sticky loops from their mycelia; the random writhing of the worms assured eventual capture by coming in contact with the loops. Soon one captured worm after another became immobile and died as a result of a poisonous secretion from the fungal strands. The fungus then penetrated each worm with mycelial threads and carried on digestion by enzymes within the bodies of the animals.

Strands of other kinds of fungi produced ring snares into which the roundworms would crawl by chance. Upon the slightest touch, the cells composing a snare would swell instantly, lessening the inner diameter and choking the worm which failed to free itself. When the soil moisture later subsided and roundworms were no longer so free to migrate, the carnivorous fungi would return to a more commonplace existence and obtain nourishment in the usual manner from organic matter in the soil.

The ways of soil fungi were many and varied, with

nearly every species having its own restrictions of food and environmental conditions. Some could break down organic material in the soil only after it had been worked on by another kind of fungus; others formed close relationships with more highly developed rooted plants. In these, the strands of mycelium either took the place of, or supplemented, the highly absorptive root hairs essential to plant nutrition. The mycelial threads provided an enormously complex network of cells which took in and conducted dissolved minerals to the roots of the larger plant. Now that water was flooding the soil, the activity of cooperative mycelia was at its height, and within each strand there was an increased flow of living protoplasm.

Among the first animals to respond to the saturation of the soil were microscopic one-celled protozoans: amoebas which crawled about carrying delicate urnlike cases into which they could retire; flagellates that would stay active until the last vestiges of water disappeared; dwarf ciliates speeding through water-filled spaces with thousands of synchronized beating cilia. Later, when the soil dried out, both ciliates and flagellates would encapsulate themselves in resistant cysts, but a house-bearing amoeba would simply retreat into its case and plug the hole, very much as the far larger land snails in the forest litter would do. Some of the protozoans fed upon organic matter in the soil, but the greatest number devoured living bacteria swarming in the soil water, swept down from the highly productive surface layers of organic debris.

Other grotesque creatures were present, scarcely larger than the soil particles they lived among. There were shiny, translucent mites with eight legs creeping carefully about; the omnipresent roundworms wriggled with increased activity now that the boundaries of their world had enlarged to a virtual infinity; water bears used their four pairs of clawed legs to wedge themselves from one cavity to the next; and rotifers unfolded beating cilia to create a fluid vortex which swept into their beating jaws bacteria and one-celled plants suspended in the water.

In the wide, humus-filled crack ·of a large boulder, a small fungus with tiny cuplike reproductive structures was being struck repeatedly by raindrops. When a drop hit the center of a cup, it splashed out countless spores that fell to the moist leaf mold some distance away. On the side of the same boulder, a red-capped lichen swelled with water and grew greener as its parasitized algae glowed through the wet fungus filaments making up the simple plant. At the base of the rock, the coiled sperm of male moss plants emerged and with long writhing flagella swam through the water film covering the bed of moss. Some of these sperm eventually would find an egg in one of the female plants, drawn to its receptacle by secretions from special surrounding cells.

From a rotten log, once nearly dried out but now a moist and pulpy mass, a slug crawled from its slime cocoon that had prevented desiccation, and inched slowly toward the ground, leaving behind a cluster of glass-clear eggs. In the same log, a red-backed salamander crept to the surface through a tunnel excavated long ago by a wood-boring beetle. At the entrance the small lungless amphibian waited with staring, protruding eyes as raindrops stirred the scene outside the log.

Another amphibian, a spadefoot toad, erupted to the surface of the soil, heaving earth out of the way as its stout body emerged from the subterranean retreat it had occupied during the past dry weeks. With small clods of mud still adhering to its yellow and black skin, the toad raised its head high, inflated a white ballooning throat, closed its eyes, and called into the night a penetrating, nasal "wank!" —to be answered at once by an identical call from a temporary puddle not far away. As the night rain continued, spadefoot toads appeared over the hillsides, met, mated in the shallow pools where females laid their eggs on plant stems, and returned to the earth, digging deep before the soil could dry out.

Beneath the soil, in a long wandering tunnel, a mole pushed its way with heavy, scooplike front feet, twisting its body so first the roof and then the sides of the burrow were packed tight. Then, turning a slow somersault in the narrow space, the mole retraced its path, stopping now and then to shake moisture from its fine, dense fur. Its deeper living quarters were completely inundated, and even the surface foraging tunnels at times were filled with water through which the mole had to pass quickly, its fur glistening like silver with trapped air. Blind though it was, it was hunting, for its other senses were keen. Normally it would hunt most actively in the daytime, when insect activity was greatest, but tonight as the soil filled with water, earthworms, grubs, millipedes, and centipedes began heading for the surface where there was a greater supply of oxygen. Their progress frequently was terminated by the mole as it scrabbled excitedly in search of victims which penetrated its burrow. Upon finding a beetle, it pressed the insect against the wall of the tunnel with one of its front feet, while it bit off the head and then began to devour the whole insect. Later, it came across a large earthworm which it dextrously manipulated until, starting at the front end, it consumed the whole length of the worm with sharp conical teeth and jaws working furiously.

In the soil under a bed of ferns were two animals which bore superficial resemblances to the mole and earthworm, but came from widely different ancestries. A dark-brown mole cricket, with powerful, enlarged front legs, shoved itself through the earth in the same manner as the much larger mammal. Water-filled tunnels had little effect upon this insect, for it carried a coat of air trapped by many short hairs distributed over its body. The other creature was a close mimic of an earthworm, but in reality was a snake, a small burrowing worm snake which, like the mole, was finding hunting more profitable than usual.

Across the forest floor small annual seed plants showed the effects of water absorption. Their delicate stalks grew turgid and straight, while their leaves flattened and were held out stiffly into the night air. Each leaf was struck repeatedly by raindrops which fragmented and splashed in a fine spray about the plants. By absorbing soil water, woody shrubs and trees were making preparations for the day to follow when the sun's energy would set into motion a vast and complex system of water transport within their trunks, ultimately releasing to the atmosphere individual water molecules by evaporation.

There was no single fate for the water which fell to earth during the night. What happened to a raindrop was determined by chance at the moment of impact. Some water seeped into the soil and flowed parallel to the surface, only to emerge as a spring farther down the hillside. Other drops continued to sink into the soil until they were trapped beneath deeply buried rock layers. Under such an impervious ceiling the filtered water would continue to flow very slowly, taking a thousand years or more until it broke free of its confinement under a distant sea.

The rain stopped in the early morning hours, but the murmuring of water dripping from millions of leaves went on, gradually diminishing. When it finally ceased, a heavy silence blanketed the forest. A single delayed drop, hesitating, plunged into a still pool below and broke the quiet with a sharp musical note.

A tiny, isolated world, replenished by the rain, consisted of a deep hole in the trunk of a beech tree, created years ago when a branch had broken off, leaving the heartwood exposed. Bacteria of decay had carved out the niche which extended down a foot or more into the thickened base of the tree. The lips to the hole were healed over with rounded bark and the cavity inside was carpeted by a spongy layer of rich organic humus. The treehole contained multitudes of organisms, populations, and interrelationships quite distinct from those found elsewhere in the vicinity. It was a world in itself.

A small black ant, having been caught far from its nest, emerged from the shelter of the treehole and crossed the damp earth in search of the obliterated scent trail it had followed before the rain. From time to time the ant paused, raised its head and body high on the last two pairs of legs and, with the front pair, scraped its antennae clear of moisture and adhering fragments of damp earth and vegetation.

A screech owl clung to the lip of another hole high in a dying oak, fluffed out its feathers and shook off a fine spray of moisture. It looked about attentively, gave its wavering, chuckling cry, and launched itself into the dark. One last nasal call of a spadefoot toad signaled the completion of its night of activity, and it joined the others of its kind beneath the wet soil. In the distance, a gray sky foretold the end of the night.

The sun rose in a cleared sky and illuminated the mist still clinging to the wet mountain slopes. Every tree and bush, all grasses and ferns, spider webs, rocks, and the soil itself, gleamed with refractive liquid jewels which caught miniature images of the sun and focused them back into the sky. Against the silence, calls of birds poured out.

Within the first hour of daylight, the sun shone through the still air with increasing warmth and intensity. Drops on every surface grew smaller as water molecules escaped through their surface film to become aerial vapor. A worker honeybee, on its way to a clover-filled clearing, landed on a raspberry leaf and sipped a water droplet. During the second hour, as drops hurried toward extinction, the glistening tracery of the earth and its vegetation abruptly vanished. A chipmunk ran the length of a fallen log, tail flicking, then sat at the end in a sunbeam, scrubbed its face and scratched vigorously.

Rain puddles lay about in profusion, but had begun to shrink. In part, this was due to further absorption by the soil, but mostly it was the result of evaporation into the increasingly receptive air. By noon only the largest pools would remain.

The enormous amount of area presented to the air by soil particles lying close to the surface allowed a continuing loss of water by evaporation. As the uppermost layers dried out, capillarity drew streams of cohesive water upward from the saturated depths, then this too left for a free existence in the atmosphere. Fully a third of the water that had fallen the previous night was in the process of returning to vapor, while a lesser amount was seeping through the soil, seeking emergence in lower lands.

With the sun pouring out radiant energy, green plants commenced their biochemical activities again, now even more accelerated by the abundance of water in their tissues. Through every seed plant, large or small, ran countless vertical streams of water, enclosed in tubes extending from the depths of the roots to the edges of leaves. Each tube, made of many dead cells and strengthened by thick walls of woody fibers, began far underground, close to regions where millions of tiny hairlike cells grew into the soil from each root tip's surface. It was only through these root hairs that water entered the plant from the soil. Short though they were, the numbers of these tiny hairs were so great even in a little plant that, if placed end to end, they would stretch many miles. In effect, each plant had miles of water-absorbing filaments in a single cubic inch of soil, making every plant of the forest highly efficient in the capturing of this precious fluid. Water entered more freely by diffusing from a region where it was abundant to another—the root hair—where it was not so concentrated. With the flow of water came all the dissolved minerals the plant would need, drawn into the root hairs by positive and negative properties of both minerals and cells.

The long conducting tubes of root and stem contained a watery sap in an unbroken column, exposed to the air at its uppermost terminal in a leaf. As the warming air encouraged evaporation, the tiny exposed surfaces of each column released a few water molecules at a time. In this way, a column of water, bound by cohesive forces, was gradually lifted by a sun engine.

For most of its journey, a column carried with it the dissolved nutrients brought in from the soil by the absorptive root hairs. These mineral foods debarked upon chemical demand in the leaves, but much of the water followed a one-way course and passed into the air from which it had come not many hours before, spraying out into the atmosphere in invisible fountains of vapor.

The water columns had been still during the night when the air was saturated, but now they moved steadily faster, drawn by the demand of twigs, leaves, and the heated air. Later in the day, when the air grew cooler and evaporation lessened, the rate of ascent would slow and even stop if the night proved to be humid.

Not all the water left the leaves, for it was needed to furnish strength by keeping cells turgid and bathed in the same aquatic environment from which all plants originated in the distant past. It also had to carry back down to stems and roots the rich concentrates of food that had been produced in the leaves by sun energy and green chlorophyll. The descent of food in small linked tubular cells just beneath the bark of leafy trees was a source of nourishment to many animals, large and small. In some of the trees, beetles were chewing their way through the bark; in others, tiny aphids were fixed in place as their needlelike mouthparts pierced an individual tube. Some of the smaller trees bore fresh scars where hungry porcupines had eaten into the nourishing tissues, or where deer had stripped bark off in long ribbons. The thick sap flowed down slowly, only two or three feet an hour, pulled down by the need of cells in the lower portions of a plant, for this was the only food of all the actively growing cells of root, trunk, and branch.

The single most important role played by water in the forest was in the chemical production of food in leaves. Here water molecules were split asunder by certain wave lengths of radiant energy from the sun, the process controlled by a large, tadpole-shaped molecule of chlorophyll. Of the two elements released by the separation of each water molecule in a cell, oxygen was set free in large quantities, but the hydrogen was passed along a series of energy traps. Later the hydrogen would be locked together with carbon dioxide molecules from the air to build a sugar that would not only supply the plant with its energy requirements, but would form the basis of food for animals that might graze upon its leaves, twigs, fruits, or strip away its bark. Food manufactured by green plants provided the basic energy for all living things, for those animals unable to eat plants directly were links in a chain of predators in which the first creatures were always vegetarians or scavengers. In this way, the water which had fallen to earth became an integral part of every living thing in the forest.

During the night, water running off the saturated earth had created rills and gullies, and filled small rocky stream beds. Further contributions to the young river would be made for days by water emerging gradually from porous stream banks. When water rushed down mountainsides and streamed across the hills, it drew upon soil water, but when the earth dried out, small streams were obliged to return water to capillary spaces of the soil, and in doing so might easily give up their very existence.

The river had been born and its growth now began. Like the elements of its diffuse creation, it still had no unity, but flowed excitedly in divergent rills and beds. Seeking lower levels, the water tumbled down, rills joining gullies, and gullies emptying into brooks. From a million separate origins, from rock funnels or meadow wetness, the river was beginning to gather and to assume the promise of a mighty identity.

IV

THE RESTLESS WATERS

A noise like of a hidden brook
In the leafy month of June,
That to the sleeping woods all night
Singeth a quiet tune.
SAMUEL TAYLOR COLERIDGE

BELOW a forested ridge, etched between the trees, a gully deepened into a rock-walled brook, shaded by overhanging birches whose pale trunks gleamed in the dark woods, catching an occasional glint of sunlight as it filtered down through the brightly lighted canopy of leaves.

Close to the stream, ferns clustered in deep crevices which split a weeping granite wall. On a narrow ledge, a black-masked wood frog sat immobile, blending with the rock and its collected humus. Only the frog's pulsating throat betrayed its presence, too slight a movement to alarm an insect victim. The frog had remained exposed since the night before, for few predators were about and there was no danger of becoming dry. Its moist brown skin caught a shaft of light and shone briefly in the dim surroundings.

Lower down the shaded rock rim, close to the turbulent water, moss grew luxuriantly, emerald and turgid. Here there was a fresh chill in the morning air, for the brook, fed by mountain springs, was very cold. The moss held on its miniature leaves a thin film of water that not only prevented the simple plants from drying out, but provided a medium in which their sperm could swim to the eggs held by each female moss stalk. The water film also was a place of existence for a community of microscopic animals: shelled amoebàs inched their way over the cells of each primitive leaflet; long-toed rotifers moved busily about and stout water bears clambered deliberately along the roughened surface.

Under an overhanging slab of the rock wall, scalelike liverworts crowded in on one another, painting the surface a deep green. Above their flat, lobed blades, a few liverworts raised small parasols composed of reproductive structures, soon to release spores for dispersal along the rocky bank.

At one point, the pitted and moss-covered trunk of an old maple lay across the brook, forming a natural bridge. In depressions of its damp surface, small carnivorous sundews grew with radiating outstretched leaves covered with sticky hairs. A tiny droplet of clear viscous fluid was held at the tip of each hair, and on two of the leaves the hairs had curled over a few trapped ants which would be digested as supplementary food.

In the precise center of the fallen tree, scales of a pine cone were scattered in disarray, evidence that a red squirrel or chipmunk had used the exposed area as a dining spot. At the base end of the log, where soil had erupted when the tree fell years ago, the wet earth held a cluster of animal prints, as the bridge was a favorite crossing point for a family of raccoons living nearby. The tree probably would not be there after another storm; on the side where it was supported by earth, the high, rushing current of the previous night had undercut all but six inches of the soil serving to hold the log.

The brook chuckled and slapped at obstructions in its

course. Each rock confused the rushing flow of the stream, causing it to veer off, whirling, in fierce eddies, spraying upward in a fine mist which glittered in shafts of light as it slowly settled over moss and liverwort, or rose wraithlike against the black rock wall. Just above a deep pool, a small torrent of white water foamed with organic matter carried from the forest floor during the storm. Rafts of foam floated across the pool, then plunged to merge in the next swift descent. Earlier, water in the brook had been tea-colored from plant residues and acids leached from the soil, but by midday the stream had cleared to its usual crystal transparency. At every bend of the brook, sticks and branches were piled high, some old and bare, some still with leaves, evidence of the night's storm. The relationship between brook and land was still one of great intimacy.

Because of the steep descent of the brook, gravity hurried the water at the swiftest pace it would ever know on land, as much as four feet a second. Its turbulence was so wild and disordered at the surface that no living thing was present, save an occasional battered insect caught in the surface tension, struggling feebly against the crushing current. Beneath the surface, tiny pebbles and glistening bubbles of air were swirled about, trapped by the swift flow after being suspended in the violence of a miniature waterfall. Only when the velocity slowed in the next pool would the little air-filled spheres rise to the surface, and the coarse sand sift down to the bottom. Each of the bubbles yielded some of its oxygen to the stream, so the water contained more of this important element than it would in its slower, warmer, downstream courses.

At times a series of thuds was followed by a dull boom as a large stone was unbalanced and rolled crashing downstream until it came to rest against a larger immovable rock outcropping. The noise under water was deafening with the banging and shifting of rocks, the loud clinks of pebbles moved about, the roar of a hundred small cascades, the musical bursts of percolating bubbles, and the abrasive rustling of sand in the less agitated pools.

The brook world grew quieter toward the bottom. Here the current was deflected by rounded, stationary stones, and its velocity lessened by the retarding effect of friction. In countless small crevices and spaces between rocks the water was nearly still and quite isolated from the storm of turbulence only inches overhead. Here there was life.

Each stone and rock sheltered a small community of animals and plants, which survived well enough, but were not tolerant of changing conditions. Some had lived in streams of this sort almost unchanged since the last Ice Age. For them, existence meant feeding and reproducing in the cold, turbulent, and highly oxygenated mountain headwaters. They even possessed sub-specialties, some preferring riffles, others pools, and still others the violence of a waterfall.

Creatures succeeding in the brook world did so largely by taking advantage of the main feature of its rapid current: the loose, scoured rocks of the bottom. Their main problem was keeping a position in the stream, which they were able to do by burrowing, clinging, or being streamlined. When a rock rolled downstream, dozens of small animals were whipped away; for a moment the water contained the writhing shapes of flatworms, mayfly and stonefly nymphs, and an assortment of other creatures before they were lost and crushed in the torrent.

Understone life was not limited to the swiftest portion of the brook, for many other small animals lived under rocks and pebbles along the brook shore. Dixa midge larvae squirmed U-shaped in the thin water film surrounding a stone above the level of the brook. Other insects, mites, and spiders, all of which breathed air, found under shoreline rocks the necessary conditions of constant temperature and humidity.

On the sheltered side of a rounded stone out in a riffle, a

stonefly nymph—a flat, spraddle-legged insect—scuttled
sidewise in search of living animal food. Its yellow-and-
black-banded body contrasted with the greenish rock, and
a series of white-tufted gills along the side of its body made
it even more conspicuous. At times it crawled to the top
of the stone where the current flowed by, but it was so flat
and had such a secure grasp on the surface that the flowing
water had little effect. Its flat body allowed it to slip into a
narrow crevice where it found a small crustacean cornered
and unable to escape, easily grasped by the stonefly. It
emerged into a quiet hollow where water swirling in pockets contained less oxygen than in the main stream. Almost
at once the stonefly nymph began rhythmically to pump its
body up and down, creating water currents which flowed
through its gills, satisfying its extreme need for ventilation
until it could return to the swiftness of the riffle.

Where two rocks nearly touched on the bottom, creating
a miniature steep-walled canyon, a fine, skillfully woven
net billowed out, tugging in the current, securely anchored
at both sides. Under one of the rocks the strands of the net
narrowed into a woven tubular case, strengthened with
small stones. At its entrance there protruded an oval,
jawed head belonging to the larva of another flying insect,
a caddis fly. From glands in its mouth it produced silken
threads and, guided by inheritance, wove the fine meshes
of a highly effective seine net.

The swift current supported little indigenous life, for it
was a transitory environment. It hurled along with it many
small living things, fragments of plants and animals swept
from the land or whisked away from under a rock displaced by the violence of the brook's flow. The caddis fly
larva's net and hundreds like it in the immediate vicinity
were constantly filled with organic debris and struggling,
dying creatures from upstream. Periodically the larva
emerged from its protective blind to sweep the net clean of
food and to repair the rents caused by larger flotsam.

An oval, streamlined shape hugged the side of a rounded
rock which projected into an especially vigorous eddy just
above the net. It could have been a symmetrical outcropping of the rock, except for its even segmentation and a
slow creeping across the face of the stone. Nothing on the
smooth outer carapace served to identify the nature of the
animal beneath, but under the broad overlapping plates,
which were a part of the external skeleton, there was the
body of an insect, the larva of a riffle beetle. This was one
of the brook's rare mobile creatures which could confront
the full violence and turbulence of the water, not only with
its strong legs and curved claws, but by a suction created as
the oval rim of its body was clamped down upon the rock
surface. While the storm of water roared outside, the insect
browsed quietly upon a fine algal coating of the stones it
crept across. As it neared a wide crevice, a stocky, hump-
shouldered mayfly nymph leaped out, braced its body with
head reared, and grasped at the oval larva with outstretched
front legs, but failed to take hold and the larva proceeded
on unaffected.

The algae were mostly green desmids and yellowish diatoms, the latter encased in boxlike shells of silica, minutely
etched and striated. In their microscopic world the current
mattered little, for the retarding effect of friction was greatest at the rock surface, and their attachment to the irregular expanse of stone was secure. Often the larger and more
permanent rock surfaces were slippery with the multitudes
of these tiny plants, which were nearly the only forms of
plant life in the rapid portions of the brook. While most of
the animals in the cold stream were predators or scavengers,
a few like the riffle beetle larva and aquatic snails grazed on
these microscopic pastures.

There were few larger animals, for size meant possible
capture by the current and this was instantly disastrous.
The bottom was irregular with a labyrinth of interconnected depressions formed between rocks and pebbles. In

these relatively quiet spaces were darters, little blunt-headed fishes. Unlike most fresh-water fishes, they lacked an internal gas-filled bladder which in ponds and lakes provides a neutral buoyancy. Darters are heavier than water and so naturally sink to the bottom of streams where the current is less and safety and food are to be found. In one crevice, a darter lay alert with protruding, black-banded eyes, elevated on its two strong front fins which served almost as legs. An insect fragment swirled momentarily a few inches away, caught in an eddy. The darter, with a flick of its muscular body, thrust out into the current, sucked in the tidbit, and returned to the safety of its depression where it sank gently to the bottom.

At the lower end of a stony apron of pebbles, the loose bed was held by a wide outcropping of bedrock. The water was shallower here and rushed over the curved edge in a smooth dark sheet streaked with shining threads of air drawn out by a heightened velocity. At the precise point where the flow was swiftest, a thick mat of aquatic moss hugged the rock, its strong stems and closely set leaflets streaming out into the current, offering minimum resistance. Within the shelter afforded by the tufts of moss, there existed paradoxically one of the quietest and safest of all brook communities. Here fragile creatures swam and crawled about in almost completely stilled water, seldom venturing up the stems where they could be caught and whirled away, lost forever to their quiet micro-world. Rotifers and water bears, similar to those in the moss along the stream banks, pursued active lives. Elongated copepod crustaceans wriggled, wormlike, through the miniature forest. Larger crustaceans, water sow bugs and scuds, lived here more safely than they could elsewhere in the open brook, where they would be eaten at once by darters, dace, and trout. Caddis fly larvae were present in the moss, carrying lightly built portable cases of leaf fragments and debris, held together by silk spun from glands. Aquatic mites and small larvae of a half-dozen different kinds were present too, moving about in their restricted world, feeding, separated from the torrent overhead by only a few millimeters of quiet water and the tenacious moss plants.

In bare areas not covered by the tufted moss, several insects of rare design had succeeded in establishing themselves. Here again were caddis fly larvae, but they were small and huddled within heavily constructed cases built of tiny pebbles, resembling miniature turtle shells. Still others had made elongated cases with ballast stones arranged lengthwise to the current. None of these larvae spun nets like their crevice-living cousins, for the constant flow of water brought directly to their domiciles all the food fragments they required, especially still living worms, crustaceans, and small insects.

Nearby, directly under the crest of the plunging water, many short cylindrical shapes crowded together anchored to the rock lip, all deflected downstream by the current. Each was a larva of the blackfly, held securely to the substrate by a disk equipped with hooks and cemented in place by a sticky secretion. Occasionally one of the larvae would loosen its grasp and in an instant disappear from the packed mass of its neighbors. Then, seeming to defy the laws of force and energy, it returned slowly upstream to its former location, a feat made possible by gathering in with a single leglike projection a lifeline of silk which had been fastened earlier to its particular location on the rock. Once there, it again attached itself, elevated its body with head forced downstream, and unfolded a pair of fan-shaped antennae which filtered from the water bits of organic matter and an occasional whole animal being swept along. Periodically the antennae were withdrawn and the accumulated food transferred to the mouth.

Beneath the rim of the rock ledge, long strands of branched filamentous algae trailed, whipping into the current. Smaller one-celled algae in the midst of the filaments were being fed upon by the small stumpy larvae of net-winged midges, which were oblivious to the force of the torrent. Each larva had a row of six suction cups running along the underside of its body, any three of which were enough to keep the insect anchored in place while it twisted its short body from side to side seeking a new position. Each suction cup had several stiff rings around it equipped with bristles and mucus, and a pistonlike plug in the center which was raised by muscles in the body, creating a partial vacuum.

How had all this life become established in areas of such high velocity and turbulence? The brook would not always be at the high level it was today, and during the preceding night it had been even higher. In late summer and fall, it might be reduced to a trickle with many of its rocks exposed and the current only a gentle interrupted flow. At these times many of the immature insects emerged into the air from their aquatic form, and adults returned to deposit eggs in likely spots. Even if the brook should not diminish, most of its aquatic insects had the ability to complete their life cycles anyway. Net-winged midges might dive directly into the water, catch hold of the bottom, and lay their eggs. Blackflies could emerge from their submerged pupal cases and be carried to the surface in bubbles of air. Moss spores, smaller even than the most minute single-celled diatoms, would catch in roughened places on the riffle crest and begin to grow.

The riffle, turbulent as it was, was the most productive part of the brook, in which highly adapted plants and animals flourished more abundantly than in adjacent pools and waterfalls. Each living thing had its own specialty of survival in the face of the torrent, and each its own mode of feeding. There was little direct competition.

In a pool below, the water grew clear and comparatively quiet. Not many plants were established here, except a few strands of the same algal filaments found in the riffle above, and occasional globular masses of small blue-green algae. The crevices common to stream bed and riffle were largely absent, for the bottom was thick with clean, washed sand that stirred and rippled in the gentle currents. The abrasive quality of particles rubbing against one another discouraged the kinds of insects and other small animals which flourished in the rushing water nearby.

The pool's surface provided a stilled habitat not available elsewhere in the brook. Close to where the riffle's last fall plunged into the pool, several broad-shouldered water striders skated into the current spreading across the quiet surface. Although the front and hind feet of one of these insects repelled the water film, dimpling it so four black shadows raced across the pool bottom, its middle pair of feet entered the water with spreading, fanlike plumes; they were its means of propulsion.

A rock slab extending straight down into the pool was encrusted with a fingerlike mass of yellow and green waving in the slow current. Flattened patches of similar consistency grew nearer the riffle area. Both were fresh-water sponges, but development into long fingers was possible only in water that did not move swiftly. Like sponges anywhere, they were inhabited by a host of small creatures thriving in the gentle food-bearing currents created by the sponge's flagellated cells. The population deep in the cavities of each sponge consisted most conspicuously of worms, crustaceans, and insects; on a still lesser scale, there were multitudes of single-celled animals, rotifers, and other microscopic forms.

Beneath the sponge, where the rock wall of the pool sloped back to form a recess, a young brook trout hovered, recently arrived from downstream. It scarcely seemed to move, but compensated for the flow of water by sensitive fanning motions of its delicate fins. From under a rock at the opposite side of the clear pool a brightly marked mayfly nymph emerged from the last riffle and wriggled out into open water, heading for shelter in a nearby crevice. In a flash the brook trout was over it, swept down and engulfed the insect with a series of rapid gulps, then swam slowly back to its station under the ledge. This was a common occurrence and in part accounted for the trout's presence so far upstream from its place of origin. In mountain streams such as this, mayfly nymphs comprised a major portion of a trout's diet. But the tables could be turned, for trout themselves were not immune to insect predators; a large dragonfly nymph edged from beneath a rock while stalking a small fingerling trout which had just entered the pool. The insect's hinged lower lip shot out, caught the tiny fish behind the head as it entered the converging focal point of the nymph's bulbous eyes, and brought it wriggling back to the sharp jaws where it was quickly consumed.

For the most part, inhabitants of the brook remained in their own localities, while the watery environment rushed on, flowing past all obstructions, obeying the command of gravity. The cold, highly oxygenated water would rise and fall, hurry or struggle to flow at all, in the weeks and seasons to come. Life contained in the brook would endure torrent, drought, and the frozen immobility of a mountain winter.

Farther down the little valley, the brook made a sharp turn around an enormous boulder that rose out of the forest floor almost to the treetops. On the other side of the rock, the brook met its twin. The second stream contributed nearly the same volume of water, but it still was tea-colored. Their confluence was marked by swirling eddies of clear and brown water which soon blended. Here floating evidence of the two streams' separate identities was trapped: leaves of birch and maple from one, and pine branches and needles from the other, for the second brook had been born at a higher altitude.

Below the great rock the stream bed grew, its musical waters now merging into a larger sound, an indeterminate gentle roar that spread throughout the valley. Far ahead a distant light suffused through the trees, a suggestion that soon the eager stream would emerge onto a grassy plain, forsaking soaked forest, cold rock beds, and a steep descent for the deliberate flowing of an increased maturity.

Water which had fallen the night before, having been carried inland from its ascent in distant warm latitudes over the southern ocean, now mingled with springs releasing water trapped in earlier rains. Water molecules were again pressing in upon one another, contained on the surface of the land, flowing downward toward their marine destination, still far away in time and miles. They had brought to the earth opportunities for highly specialized life which had to be surrounded by fluid water, using it in the construction of tissues, for the manufacture and utilization of food, and for reproduction. The river had come into being and was growing with the exuberance and mighty surge of youth.

V

OF MURMURING
STREAMS

By shallow rivers to whose falls
Melodious birds sing madrigals. . . .
CHRISTOPHER MARLOWE

T H E young river left behind the dark mountain slopes and emerged onto a bright upland plain, its widened course fringed by shading willows and alders. It passed between flowered banks where afternoon sunlight flooded across to spangle the slowly churning water with a thousand miniature suns.

With a harsh, rattling cry, a kingfisher plummeted from a cluster of oaks on a high knoll, fluttered to a halt in mid-flight twenty feet over the water, and plunged straight down, striking the water with loud impact. Its heavy bill cleaved the surface and snapped up a young dace sunning itself in the shallows. The bird, with powerful strokes of its short wings, then rocketed upward to the next grove downstream where it alighted on a bare branch overhanging the water. With two sharp blows, it pounded the fish to quiet its struggles, and quickly swallowed it headfirst. There the kingfisher crouched on stubby legs, its huge

crested head cocked for a telltale movement of another fish
far below.

This was familiar territory to the bird. Earlier in the year
it had excavated a long tunnel into the stream bank, directly under the tree in which it sat. Fifteen feet into the
burrow its mate had laid a half-dozen eggs, and both the
parents had supplied the hatchlings with an endless supply
of minnows, dace, shiners, and other small fishes from the
stream. Now the burrow was empty, but its floor was
littered with myriad fine bones of past meals.

A large green darner dragonfly flicked across a nearby
open field to the stream margin, where several lesser
amber-wing dragonflies scattered to the shelter of a stand of
broadleaf arrowhead growing in quiet water behind a large
shoreline rock. The great insect looped out over the water
and back, hovering frequently, its bulbous green eyes scan-
ning every flying movement within several yards. After it
darted away, the amber-wings held to the arrowheads
briefly, then began their coursing up and down the stream
shore, spaced apart facing the water, for this was their time
of mating and each defended its own territory from others
of its kind.

Except in the shallows, the bottom was no longer distinct. The stream had grown turbid with sediment washed
from thousands of temporary drainages created during the
deluge of two nights before; only a very small amount of
the suspended cargo carried by the stream was derived
from its own bed and shores. The gently sloping valley
floor over which the young river ran was composed of silt
and sand particles much like those carried by the water, for
these in fact were the source of the valley plain's growth.
The stream bed was only occasionally studded by larger
rocks.

Light which filtered downward through the stream,
winking in ever-shifting patterns, was yellowish and lacked
the blue coolness of the summer sky. The moving water

appeared permanently murky, but the course of each tiny suspended particle did not follow the apparent smooth flowing of the stream. Grains falling to the bottom were soon picked up again by the current, where they were thrust gyrating, bobbing up and down, among forces that carried them downstream. Some were trapped in eddies along the bottom; others were injected into a central thread of high-velocity water for a hundred yards before falling out again. In quiet spots they descended in a soft rain which blanketed the bottom with a fine, loose deposit. There the bottom sediment would remain until vagaries of the current or an increase in velocity would cause its particles to quiver, pulsate, and be lifted once more into the swirling stream overhead.

In swifter portions, the rocks and coarse sand of the stream bed were restless and were pushed or tumbled along by erratic currents roiling between the high-speed jet running through the center of the stream and the friction-held quiet water pressed against the stream bed. These swirling thrusts at times would reach down and with forceful impact dislodge a rock and send it rolling downstream. Some stones, too flat to roll, slid or bounced along, stopping and starting, caught between two opponents—the persistent current and the rough bottom, jagged with obstructions. The constant dashing of rock against rock, the jostling, pummeling, random collisions, broke off corners and rubbed them smooth. Fragments were ground into rounded shapes, although the finest sand particles remained irregular, for the water itself kept them from violent impacts against one another. An occasional stone, trapped in a rock-bound depression, scoured out a deep pothole while it was turned into an almost perfect sphere.

In one spot, an outcropping of rock greatly narrowed the stream, creating a funnel only a few yards wide. As water approached its swift descent through the funnel, a general restlessness grew across the stream bed until there was a stormy transport of everything on the bottom save the largest rocks. Sand grains vaulted off the bed and vanished at once. Below the funnel, where the current grew gentle once more, stones bounced to a halt and the smallest grains settled at last to the bottom far downstream. The swift turbulence in the funnel created a cloud of bubbles which lightened the water and, spreading out, broke with noisy effervescence in the quieter reaches below. Twigs, leaves, and other inanimate objects shot through the gap, but there were few living things that were displaced, for most of the bottom creatures immediately upstream dwelt securely in tubes or under rocks.

On the surface above the narrowed spillway, two lines were etched across the water toward either shore—lines that revealed the passage of rapidly moving water flowing past two wedges of still water held against the rock dam. In the wedges, flotsam had collected, and water striders darted over the surface. Below the streamlines, anchored clusters of simple, filamentous green algae whipped out from rock surfaces.

Across the stream bed different algae lay upon flat stones in encrusting patches; some were red, while others were brown or blackish-blue and mucilaginous. The motile spores which had given rise to these simple plants, crusts and filaments alike, had been able to attach earlier to a rock surface where the current lacked strength, retarded by friction. Often the upstream face of a rock, around which the stream flow was diverted and slowed, became blanketed with low-lying growths, as was the lee side with its confused and reversed currents. The sides of most rocks, where the flow was compressed and passed by swiftly, were bare of plants of any sort.

Conditions in the stream encouraged plant growth, for the water was rich in dissolved nutrients and gases contributed by land and air. Both were mixed thoroughly in the water by turbulence. Wherever the current permitted, plants of all sizes and kinds flourished. Many of the hair-thin filaments of algae were heavily coated with still smaller plants glued singly or in bunches. Among the clusters of these one-celled, stalked diatoms, even more minute green cells swam with long flexing whips through an undisturbed world while the stream rushed by outside.

Corresponding microscopic plant cells swam or rested in sedimentary accumulations on the bottom. Some were geometric chains of precise symmetry; others were radial colonies of pointed green cells, kept from sinking in the soft silt by their extensive surface area. Here they were grazed upon by very small, energetic crustaceans similar to those in the moss of the mountain brook. These elongated, specialized creatures were the only representatives present in the stream of the vast copepod populations that swam in surface waters of the broad estuary two hundred miles to the southeast.

In the center of the stream, a full-grown brook trout hovered over the bottom, its mottled, olive-brown body almost perfectly matching the graveled bed. This trout was far larger than its young relatives in the mountain headwaters. With just enough force to keep it in position pointing upstream, the fish's muscular body undulated slowly. The trout, like most of the stream's inhabitants, large or small, always faced the current, for it was from this direction that food and possible physical harm might come. To be swept out of control downstream meant almost certain danger and death. A sudden, unusual movement underwater or from the world beyond invariably caused the fish to rush upstream, unleashing powerful muscular contractions into sinuous bends which thrust laterally against the surrounding water, a propulsive force at once simple and enormously effective.

The brook trout waited in position, its senses tuned to the moving, noisy world of the stream. A long, sensory canal extending down each side of its body picked up dull, low-frequency sounds and pressure changes from the water, while its ears, located deep within its head, listened for high-pitched noises that might indicate food. Its U-shaped nostrils continuously sampled the stream flow for dissolved odors.

The trout's eyes flicked in unison back and forth, now downward, now to the fore. They were linked in almost toggle fashion by nerve and muscle, and could not converge to see beyond the nose; only one eye at a time could be directed forward. Nevertheless, the trout's field of vision extended upward as well as forward and to the rear, and its vision was keen.

One of the amber-wing dragonflies ventured out farther than usual during its territorial flight, and hovered inches above the surface where the trout lay. As it darted toward the shore again, the water beneath it erupted and the trout's brown, glistening body was for a moment air-borne. The foot-long fish splashed down, and the air was empty. In one great burst of coordinated energy, the trout had propelled itself off the bottom, adjusted to the rush of the high-velocity jet current in the water overhead, compensated for the difference in refraction between air and water, determined the course and speed of the insect and its distance above the surface, and returned to the safety and obscurity of the stream bed.

For a predator like the trout, the stream was amply stocked. In the last hour, it had eaten several dozen midge larvae, many small crustaceans, several snails, a large leech, and a number of young fishes, including a few of its own kind. This was the brook trout's world, and one which it largely dominated. While it might ascend farther upstream, this particular kind of trout was not found in the great languid river in the wide valley to the east, for it required the aerated, cool waters of upland streams and was quite incapable of surviving in regions where life-giving oxygen was dissolved in lesser quantities.

Soon, as the summer drew to a close, the trout would seek a mate, as it had the year before, and the elaborate ritual of nest-building, courtship, and spawning would commence once again. The pattern was always the same. First, the female would descend to the bottom, lie on her side, and wash out a depression in the loose gravel of the stream bed with vigorous thrusts of her body and tail fin. She would do this repeatedly until the nest was several inches deep, all the while being watched by the guardian male hovering nearby. His function at this time would be to drive off other males should they attempt to approach the nest. Later, he would court her by swimming overhead, butting her with his snout, and quivering rapidly in a rehearsal of the mating act. When the nest was complete, the female would lie in it on her side, the male pressing close beside her. With bodies vibrating in quick shimmering movements, they would simultaneously shed eggs and sperm in a pinkish cloud. It was done in only a few seconds, and the male would then retreat. At once the female would sweep gravel from the edges of the pit over the eggs with her anal fin, heaping them into a protective mound. Fertilization and covering were accomplished so quickly the current would have little chance to wash the sinking eggs away. Soon the eggs would absorb water and become turgid; within days a tiny embryo would be visible, although it might be three months until the young could hatch, struggle through the gravel bed, and enter the open stream. After each frenzy of nest-building and mating, the parent trout would repeat the process in a nearby area of the stream bed, until the female was finally rid of all her eggs several days later.

All this the brook trout had done for the first time the previous year, instinctively and untaught. The time was drawing near again, and the fish was eating voraciously, for during the period of spawning all feeding would be suspended.

The entire stream had a very different character than it had possessed earlier as a mountain brook. There, with its steep descent, it had been narrow, often deep, and frigid. But in the valley, many miles and several days later, the enlarged stream was descending far more gradually and had widened within a shallow bed. The slow diminishing of slope would continue until eventually it would descend vertically only one foot in a mile.

The brook's course had been so turbulent that the water it sent cascading down hillsides had been unlayered and thoroughly mixed, but the stream now had three distinct zones of water: its thread of high velocity which usually ran down the center, but often veered to one side or the other; a quiet region close to the stream bed; and a restless layer of agitated water between the two. Were it not for friction between water and stream bed, and friction between the different layers of water, the fluid stream might in theory flow over a thousand miles in an hour, yet the retarding forces were so effective that a water particle streaming across the upland plain, pulled along by gravity, would do well to cover two or three miles in the same length of time.

With a lessening of current close to the shore, and an accumulation of rich sediments, rooted plants were able to take hold and grow upward through the water, reaching the surface, where their stems would continue another foot or two into the air before producing leaves and flowers. The most common of these was broadleaf arrowhead, which provided not only refuge for the amber-wing dragonflies but food for muskrats and a variety of birds. Many of the plants were displaying tall, whorled spikes of purple flowers that were being visited by bees and other insects.

Different plants, without the ability to stand erect in the air, filled a few of the quiet bends with floating leaves. The surface leaf of river weed was broad and buoyant, but other leaves the plant produced beneath the water were long and narrow, offering little resistance to the current. In areas where river weed was common, its pliable stems provided a place of attachment for many small cylindrical animals, brown hydras whose delicate tentacles streamed out in the current. Microscopic stinging cells on the tentacles enabled a hydra to capture and paralyze a smaller creature floating by, whereupon the tentacles would contract toward a conical mouth, gaping in anticipation. Larger snails and insect larvae crept along the waving stems, not harming the multitudes of tiny one-celled vorticellids that jerked out of their way on contractile stalks, or retreated into urnlike containers.

By now the sun was low over the hills holding the little valley, and the sharply angled light was warm and golden. The frigid constancy of the mountain brook had vanished long before, and the cool meadow stream increasingly resembled the mild summer air.

Dead leaves floating down the stream caught the light and were transformed into glowing luminescence. A single aerial seed came riding down the surface film, its fine, gossamer filaments radiating outward like a brilliant, crystal star. Just above a riffle area, several dozen elongated water striders kept their position in the current, veering near one another, feinting, investigating flotsam particles streaming by, hopping over floating leaves, and maintaining an activity equal to the speed of the stream. Occasionally, with frenzied spurts, several would take new positions upstream. Like their relatives of the mountain brook, they were supported on the surface film by front and back pairs of legs which bore water-repellent pads and hairs. With every thrust of their middle legs, the insects skittered forward into the current. The tiny depressions in the stream's surface from a water strider's power stroke left behind two instantaneous flashes of reflected sunlight that quickly

winked out. Across the stream these tiny, explosive bursts
of light revealed the presence and activity of the busy,
scavenging insects.

At times, water striders scattered widely as a crinkled,
dried leaf, riding high on the water, caught a gust of wind
and, defying the force of the current, sailed spiraling upstream. Along the banks, eddies in the wind carried many
such leaves along until the current captured them again and
reversed their course.

In the last minutes of direct sunlight, two blue jays cried
incessantly from a shoreline willow, and a flock of red-
winged blackbirds, with flutelike cries, flew from one burr
reed to the next, the males displaying their brilliant shoulder patches. An unseen phoebe called briefly, and a pair of
cedar waxwings flew low against the darkening water. As
shadows crossed the stream and its valley, most bird voices
were stilled.

With the cessation of sounds of the day, the stream's
liquid noises seemed louder and filled the quiet air more
completely. New voices arose, predominantly the guttural
snores of meadow frogs from the water's edge. As a male
frog prepared to sing, it inflated air sacs along either side of
its lower jaw to serve as resonating chambers when air was
expelled forcibly from its lungs. Each male lay spread-
legged in the water, only its head visible, buoyed by the
yellow air sacs.

In a dark tree overhead, a black-crowned night heron
walked out along a leafless branch, methodically placing
one splayed foot after the other. At the end it stood motionless, hunched compactly, taking in the dimming world
beneath, until with an unmusical squawk it launched itself
over the water in search of one of the meadow frogs.

The night heron and the frogs were only two of many
kinds of animals tied to the presence of the stream, day or
night, for breeding, shelter, or food. The water molecules
that had awakened an air-borne bacterium, that had been

absorbed by seeds and spores and had created a miniature soil-water world for tiny worms and protozoans, now were immensely concentrated. The energy imparted to them during evaporation from the ocean was being released during their gradual descent along the stream channel. Water molecules and their pathways were the keys to millions of interrelated lives in the small river valley. The heron and a few other large animals reposed at the apex of existence here, a pyramid built upon the production of food by plants and its consumption by animals. All these living things were extraordinarily complex, yet each was composed largely of water that had to be replaced continually, each required water in its biochemical machinery, and each was tied to this aquatic environment for a way of life. Every fish and insect, frog or alga, maintained itself by means of the simple, symmetrical molecule possessed by the stream in such abundance.

For animals of the night, day was a time of comparative inactivity; as the world darkened, a whole new series of pursuits commenced in and around the stream. Midges emerged from an aquatic larval state during these crepuscular hours, dancing over the water in swaying vertical columns. A bat zigzagged across the valley and with acrobatic frenzy dashed repeatedly through the column of midges, its bursts of echo-locating cries far beyond the hearing of any other ears.

Although the water was now quite dark, the nature of the stream varied not at all. There was no change in temperature, nor in the quantities of organic bits and pieces it carried along. The animals that could make use of this potential drifting food were as active now as they had been in the daylight, for one time was as good as the next for filtering material from the undiminished supply.

Irregular encrusting masses of living sponge grew upon the rocky bottom, or attached to a plant stem. The loose association of cells comprising a sponge were held together by vast numbers of needlelike and three-pointed spicules of a glassy substance. These fresh-water sponges lacked great size, but like their marine relatives they were efficient filter feeders. Internal currents created by multitudes of flagellated cells within a sponge flowed through a maze of interconnected chambers and channels before emptying back into the watery environment. Each sponge in the stream was capturing bacteria and minute organic particles that came into contact with delicate collars surrounding every flagellated cell; there they stuck, were absorbed, and were passed on to other internal cells for digestion.

In daylight, some of the sponges were a bright green rather than their usual brownish yellow. The color was caused by minute green algal cells taken in by a sponge during its normal feeding process. Many of the algal cells were not digested at once, but were able to multiply, their presence almost invariably resulting in better sponge growth. Now that night had come, some of the plant cells were being digested by the sponge, but not enough to diminish the green color by morning. Over a period of many dark and cloudy days, the color might indeed fade. Sponges on the undersides of rocks were not green, for the algal cells were unable to reproduce effectively and were soon digested.

The sponges in the stream were healthy on this summer night, but later in the fall their tissues would weaken and eventually disintegrate, releasing masses of the geometric spicules to the water as well as many tiny yellow spheres. Because a fresh-water sponge could not survive the rigors of winter, it had to have a means of producing new colonies the following spring. Each fall it gathered together groups of unspecialized cells, covering them with protective membranes and a honeycomblike crust. These capsules waited dormant in a crevice where they happened to lodge, perhaps even frozen in ice. Cold weather seemed to be necessary to set a living clock into action in a capsule, for without such exposure it seldom germinated in the spring. When the water finally grew warmer, a capsule's shell would crack, cells would creep out and cover it, and soon a new young sponge would grow outward from the crevice.

Filter feeding was an effective way of life in the stream. Throughout the finer gravel and sand there were many such animals, including tiny pill clams, no larger than a pea and almost as spherical. Like their much larger marine relatives, they stretched out two siphons, one to take in water and one to discharge it. A constant flow brought in particles of organic matter that became trapped in sheets of mucus on the layered gills, and were then conveyed to the mouth by thousands of tiny beating hairs. The gills were versatile structures; not only did they take in oxygen and food, but they were greatly distended by enclosing almost two dozen young clams. The gills, in fact, were marsupial pouches. At present the young were very small, but at the time of their release later on, each would be almost a quarter of the parent's length. They could fit within the older clam's shell only because its valves were highly arched and almost formed a sphere.

Pill clams were hardy and adaptable to all sorts of stream conditions, and even had been carried to entirely new locations encrusted in mud adhering to the feet of water birds. This was not an unusual mode of transport, for snails, sponge capsules, and the eggs of many aquatic creatures— even certain fishes—had entered the stream or been carried away to other streams and ponds on the feet of killdeer and similar wading birds. Still other eggs and dormant bodies were capable of surviving a trip through the digestive tracts of ducks, and so had a chance of being transported to new and favorable aquatic situations.

In the greater darkness of crevices, overhanging rocks, and burrows leading back into the bank, there were stirrings of hidden inhabitants. The first sign of a creature in one of these retreats was a pair of long, slender antennae that flicked about with nervous sensitivity. Next to appear was a pair of stout, tubercled claws, followed shortly by a pair of black, hemispherical eyes borne on stalks. Soon a complete crayfish emerged from its daytime hiding place, walking forth daintily, poised on four pairs of legs, its abdomen held arched with tail fan outstretched. Almost at once the front two pairs of pincered walking legs began to probe in the sediment, picking up bits and pieces of decaying matter and conveying them to the intricate mouth parts. The big claws tested and tore at larger objects. Most of this activity was in the vicinity of emerging vegetation along the shoreline, where there was plenty to eat and the passage of water was not so swift. Scavenging for organic remains here was profitable, not only for crayfish but for other smaller crustacean relatives, scuds and isopods.

A crayfish seldom came in contact with other living creatures, but when it did and made an attempt at capture, it usually met with little success. Decaying aquatic vegetation comprised the major part of the crustacean's diet; animals, even the small and defenseless dwellers of soft bottom sediment, were unimportant items in the total volume of food consumed.

During the daytime, the stream's crayfish had remained in their burrows, under stones, or in small depressions, their rounded eyes colored an opaque brown like the rest of their bodies. At dusk, when light intensity fell, a gland in the stalk of each eye was affected and it in turn directed the hemispherical eyes to retract their pigment and become far more efficient gatherers of light. As a result, in the evening their eyes turned from a brown to a deep and penetrating black. The little glands, despite their peculiar location, controlled a variety of other important functions in the crayfish as well, especially growth and the periodic molting of the outer skeletal armor.

One of the crayfish ventured from the protection of the underwater forest and crawled toward the depression where the trout lay. Its movements were revealed to the trout by the last remaining vestiges of light in the sky, and the fish made an abrupt lunge at the crustacean. The crayfish, reacting violently, doubled its muscular abdomen and tail fan with a powerful, convulsive thrust. It shot off the bottom backwards, but not before the trout's jaws snapped closed on two of the walking legs. The escape of the crayfish was hardly affected. At the bases of these two legs, underneath a double crease in the outer skeleton, special muscles contracted in an instantaneous reflex, severing nerves, muscles, and other tissues. The legs broke off at once and the crayfish was free. It continued to pump its abdomen virogously, until it was securely wedged within its familiar crevice. The brook trout swallowed the two legs and glided back to its own depression out in the stream.

Later, after the next molting of the crayfish's outer skeleton, two tiny legs would be revealed in place of the missing ones, and with each successive molt the legs would grow larger until they finally would be indistinguishable from all the rest.

On this summer night, feeding and defense of territory were the primary activities of crayfish. In other seasons, reproduction would be added. Earlier in the year and again later in the fall, each male would develop a pair of special reproductive appendages for the transfer of long strands of sperm to the female. In summer, after one molt and before another, these same appendages lacked the capacity for sperm transfer and reproduction was impossible. Breeding had occurred extensively in the vegetated area of the stream some weeks before; now most of the females were bearing hundreds of developing eggs under their bodies, each egg attached securely by a short stalk to abdominal appendages that fanned back and forth, ventilating the egg mass as a whole. After mating, the females had held the cords of sperm for days until their eggs were ready, and only then did fertilization occur. Each female had lain on her back, released a few eggs at a time, at which time the plug to the sperm cord broke, the sperm flooded out, and several new lives began. The eggs had developed rapidly since then, already showing evidence of crayfish form: each had two minute black eyes, a humped body, and tiny delicate legs that waved feebly.

Crayfish, seldom seen during the day, nevertheless were fed upon by fishes of all kinds, raccoons, otters, snakes, turtles, frogs, and wading birds. They were vulnerable once their hiding places were disturbed and burrows penetrated. What made crayfish important links in the living economy of the stream was their conversion of decaying organic matter into their own substance, thus making energy and essential ingredients available to predators: the rich deposits lying in quiet backwaters were once more returned to the chain of living things that existed near the stream or within it. Furthermore, crayfish were parasitized and lived upon by a variety of smaller animals. Tiny two-shelled crustaceans crept across their gills, as did large numbers of a minute, leechlike segmented worm; long hair-like worms grew within their bodies. The outer skeleton of a crayfish that had not molted for some weeks had attached to it multitudes of single-celled animals and diatoms, and even filaments of algae.

Crayfish were good burrowers in the stream bank, but they lived only a year or two, so many of the holes were now occupied by other animals. In several burrows along one bank, salamanders began to emerge for their night of hunting. A mud salamander paused at the entrance to a tunnel a little above water level; there it remained with bulging, staring eyes and pulsating throat. It would have been inconspicuous under any conditions, although the young of its kind were obvious in daylight with their bright colors and dramatic markings. When it was young and in a larval state, the salamander had dwelt within the stream, taking in oxygen through bushy, blood-red gills that stood out from its head, and it had spent much of its time under flat shoreline stones. Other burrows nearby had been enlarged and were inhabited by water snakes, which not only fed upon the salamanders but ventured out into the stream to capture crayfish, frogs, and small or weakened fishes.

As the evening wore on, a glow behind the forested hills foretold a brilliant moonrise. There were few clouds, and before long the night world was flooded by a cool and silvery light. As the sensitive eyes of stream animals were now able to pick out their surroundings clearly, some of the activity diminished, for there was increased danger from predators.

Several shrill, chirping calls interrupted the gentle snoring of frogs and the insect voices of the meadow. They were followed by a grunt and chattering; from nearby, a series of chuckling sounds replied. The grass above the shoreline parted and a broad, flat head looked out. With only a moment's hesitation, a graceful and darkly glistening otter plunged over the brow of the bank, pushed off with all four feet, and slid down the clay incline. She was immediately followed by three smaller, elongated shapes—her pups. Now that moonlight filled the stream, the female otter used her eyes to search for food, although her stiff whiskers also aided in detecting prey in the slightly turbid, churning water. Within seconds she found a crayfish. Along the shore where the arrowheads stood there was a frenzy of food-seeking by all the otters; for a while they were intent solely on eating crayfish, but soon the young tired of this and began a skillful game of follow-the-leader, bounding and twisting through the shallows. Their smooth, rolling play in the water was extraordinarily graceful and hardly revealed the high degree of coordination necessary. The mother joined them briefly, then with a sharp bark, she swam toward shore and was immediately followed by the pups. They climbed the slippery bank with curious humping motions and gained the grassy meadow, where each in turn stopped and shook its coat vigorously, sending a spray of water shimmering in the moonlight. The young otters rolled over and over in the meadow, while the female repeatedly wiped her face on the grass. The pups continued their boisterous play, tumbling about in mock ferocity, occasionally crying out with a sharp bark when the play grew too rough.

By now, this area of the stream was thoroughly awakened to the otters' presence, so hunting no longer would be very profitable. Once more the female slid down the bank, calling her pups to follow her, and all swam directly upstream. The female, in the lead, almost cornered a brook trout between two rocks, but it flashed off and she abandoned its pursuit, continuing her smooth rolling dives toward a pool she visited regularly. All this was thoroughly familiar territory and marked against intrusions by other otters. Wherever she had gone ashore, she had left bits of twisted vegetation and a few drops of scent to post the area as hers.

Amid the willful and unpredictable activity of the otter, trout, and crayfish, there was one environmental constant: the flow of the stream. It was persistent, seeking lower elevations always, no matter how tortuous the course. Even the slightest drop in the stream bed assured a continuance of flow, but it followed an intricate path, many times longer than the shortest route to the sea. The young river never flowed straight down the valley, but curved and bent back upon itself in an irregular and always-changing pattern. In places where a tributary entered the main stream, their confluence forced a change in direction and the stream was diverted in response to their respective and contending forces. Elsewhere, the stream would only flow a hundred yards or less before bending sharply. Curving meanders were inevitable; it was physically impossible for the stream to flow for any length of time or distance before assuming a sinuous path of ever-tightening loops.

The deepest part of the channel did not remain in the center, but wandered from side to side within the stream bed, close to one bank, then crossing to the other. Where the greatest volume and velocity of water pressed against the bank, it undermined the soft earth, increasing the width of the stream at the bend. On the outside of one such curve, the bank tended to be steep, with the root systems of trees partly exposed as they turned to grow back into the loosening earth. Sections of the bank had collapsed, but the quantities of soil entering the stream had been quickly carried away. Trees and large rocks, undercut and without support, had fallen into the water where they fortified the steep wall of the outer bend against the tremendous pressure which swept around the curve with greatly increased centrifugal force, shearing and carrying away everything unable to withstand its power. Here the water moved at high speed and more smoothly than in shallower portions of the inner bend.

How had such a meander begun? Perhaps its birth had been a minor and insignificant affair: a small rock or log protruding into the river on one side had diverted the flow slightly from its straight streamlines. This created not only an eddy, but an obliquely flowing current that wandered across to the other bank where it was reflected back in languid fashion toward the first shore, but far downstream. Once this slow zigzagging commenced, forces were set into action which eroded first one bank and then the other, the river always seeking to lessen the work being done by distributing it down the length of the stream bed. As undercutting and erosion deepened, the forces were intensified by the pressure of flowing water and a meander was born.

During the night, an old oak that had stood on the valley

floor for a hundred years began to shift slightly as the outer curve of a meander cut away at the steep bank. In the early morning hours there was a vague upheaval in the sandy soil surrounding the oak. There were muffled splintering cracks beneath the soil surface and slowly, very slowly, the tree tipped toward the water below. Its thick foliage described an arc against the lightening sky, accelerating as it fell. Soil on the uphill side of the tree suddenly erupted, spewing clods of moist earth and stones in a ten-foot radius. With a tremendous report, one major root snapped in two, then others appeared above the soil, some thoroughly rotted out. The oak swept down, splintering saplings and branches off at their bases, until with a mighty roar the upper third of the tree struck the water, sending thousands of separate momentary fountains into the air where each branch and twig struck and submerged. Small slides of damp soil tumbled down the bank and were whisked away downstream. The great cavity left by the broken root system revealed raw, yellow earth, very different from the rich black humus which lay across the flood plain. The whole tree trembled as the current tugged at branches and leaves, whirling away those that were fragile or broken.

A meander created several distinct environments, each of which supported its own representative plant and animal populations. In an outer bend where water was deep and so swift that sediment was unable to accumulate, few living things could withstand either the current's massive power or the shifting, unstable bottom. Swimmers as vigorous as the brook trout could pass through, but most mobile animals avoided the region and sought the lessened velocity of an inner bend. On the inner convex side of a meander, where the slope was more gradual and the current gentle, sediment and small cobblestones, deposited in humps and low bars, had been brought down from the eroded curve directly upstream. The exposed cobbles along the low inner bank revealed a continuing source of stones that had been deposited by a much larger stream during past ages when glacial sheets were melting and producing great rivers in every valley. The rounded cobbles, all of uniform size, extended downstream in a long tongue, creating a loose and permeable breakwater; on the inner side lay quiet water supplied by a slow seepage through the stones. It was, in effect, a tiny pond almost a foot lower than the water streaming by outside the stony wall. On the outer cobbles the usual stream insects lived: stoneflies scuttling over rocks, and caddis fly larvae with their silken nets spun to trap organic flotsam. In the breakwater itself aquatic earthworms inched through the tightest cracks, consuming the organic matter trapped there. Along the whole length of the rock tongue, wherever the stones were moist, slowly waxing and waning points of light identified creeping glowworms, the predatory larvae of firefly beetles that were tracing rising arcs of light over the meadow in the valley. Submerged stones in the pool behind the tongue were covered with a brownish coat of one-celled diatoms. Narrow, wandering paths through the velvet coating ended with a snail rasping off the microscopic plants with its filelike tongue. Overhead, on the surface, a miniature species of water strider, more fragile than the large ones out in the stream, skated back and forth, interrupted at times by the rapid coursing of whirligig beetles. The beetles etched behind them sharp, silvery lines of reflected moonlight across the black water. Spider webs connected many of the emergent rocks along the edge of the backwater; most of the webs consisted of small snares and were not very complex.

The tongue itself was the result of erosion from upstream, where water cutting out the concave side of a meander produced a steady flow of pebbles and sediment down the same side of the stream. It was deposited along the inner curve of the next bend, tending to build out ever farther.

At the end of the tongue, cobbles gave way to sand and silt. In the dark mud were many little pits, evidence of the daytime activity of mud-dauber wasps which had scooped up pellets of mud to use in the construction of their nests. Although the muddy sand was coarse with mica flakes that glinted in the moonlight, it held accurate impressions of tracks made recently by animals, mostly raccoons during their search for crayfish, molluscs, or perhaps even a fish sleeping in quiet water. Another set of tracks was evidence that a deer had walked down the tongue to the silted end to drink, and had then bounded across the pool to the young forest which stretched to the river bank.

The current spun with great speed against the outer bank of each meander and veered across the stream, continuing to whip back and forth in a series of eroding pressure points that created tight curves down the whole length of the flood plain. As the years had passed, the sinuous course of the stream constantly took on new patterns; the successive growth of vegetation on the flood plain followed the outlines of old stream beds superimposed upon one another. There was a high degree of regularity to the system of meanders, for the stream sought to minimize its work of erosion over the valley floor. Nowhere did it run straight for a length over ten times its own width. Over a period of time the meanders tended to shift across the narrow plain, creeping laterally until opposed by the barrier hills.

A meander might become so tightly curved that a new channel would break through the restraining neck of land, suddenly creating an island. The main flow avoided the old loop, and soon deposits of sand and silt, dropping into the quiet water, walled it off with an effective dam, isolating it from the main channel. In one spot along the flood plain such an ox-bow lake, a shallow crescent-shaped pond, had been formed a few years before. Tonight its still surface reflected the full moon as a distorted disk whenever a muted ripple passed by.

The stream hurried on, descending through the ever-widening valley of the rock-floored piedmont toward the vast coastal plain below. No longer was it in danger of disappearing into a swamp or losing its contents to the absorbent soil. Now it was a shaper of the land, cutting through, displacing, or avoiding all obstacles. It added to its volume with every mile as smaller hillside streams entered and were carried off, their separate identities lost in the blending. The stream's voice had changed as well. It did not splash and sing with youthful exuberance, but in its murmuring there was the promise of a deep roar of power. The growing river gleamed in the moonlight as a road of silver.

VI

THE RIVER'S CHILD

*A lake is the landscape's most
beautiful and expressive feature. It
is earth's eye. . . .*

HENRY DAVID THOREAU

Isolated from its parent stream, the crescent lake
lay wrapped in a mantle of rising vapor, its surface un-
marred in the early morning stillness. The air had grown
cool during the night, but the small lake held warmth from
long sunny days. In the hours before dawn, water molecules from the surface had been escaping their bonds of
hydrogen linking them to their neighbors. They rose steadily from the pond, condensing in the chilly air, gathering in
wraithlike billows swaying together or pulled apart by the
gentlest of air currents. With the sun climbing beyond the
blue hills, evaporative activity would soon increase until
the last wisps of vapor faded. No matter how hot the day,
the level of the pond would remain constant, for the dikes
of sediment sealing both ends allowed a slow seepage of
stream water to enter the shallow basin.

Fog steaming from the pond blanketed much of the
valley and settled in swelling droplets on every stem and
blade. It was one means by which water molecules could

rejoin their river, for they condensed so heavily along the banks that tiny rivulets coursed into the running water. Bright dewy webs stretched against the gray sky; rocks bared in the meadow grass gleamed darkly. While salamanders still hunted along the wet banks, the banjo-plunks of green frogs and sharp clicks of cricket frogs grew silent one by one as the eastern sky lightened. There was a last deep and sonorous roar from a bullfrog as the day began. The air still held too much moisture for most daytime insects to fly, but redwing blackbirds stirred in the cattails, with shrill whirring calls and a fluttering of bright wings. A lone little blue heron stood silent on the shadowed shore, yellow eyes gleaming, with bill poised above a pickerel frog lying quietly at the surface.

In the small lake the intimacy of water and life was clearly evident. Water swelled the tissues of larger plants and animals, suffused cells and giant molecules, and played a role in the invisible world of atoms and atomic fragments. At least two-thirds of every living thing in the pond or on its shores was composed of water.

Fluid water existed in cells, the structural units of all life, in tiny vacuoles or in protoplasm itself, but it also dwelt more importantly as a part of the many intricate membranes of a cell. There were membranes which sealed every cell as a unit; others wove through the substance of a cell and its nucleus, and some were even found inside of the smallest components used in the transformation of food energy or the growth of the cell. Nearly all major life activities of a cell took place on the surface of these membranes.

Each membrane, part protein and part a fatty material, held thin layers of water molecules arranged in a pattern resembling the structure of ice. The resulting layered membrane, no matter where it was in a cell, had distinct and remarkable properties of allowing certain substances to pass through, while refusing passage to others. Each cell then had an ability to regulate its existence by satisfying needs and eliminating products and wastes. Such an orderly molecular lattice influenced nearly all of the basic activities of every cell, helping it to adjust its life pattern to changing conditions of the environment, although life could not persist at environmental extremes. If they grew too hot, proteins would coagulate, and freezing in winter could destroy the delicate membranes and their vital functions.

The latticed water molecules of a membrane were unquestionably essential to the basic structure of life and to the passage of materials, yet water in a cell mediated its internal activities in still another way. Every living cell was in a state of constant activity, whether it could be seen to move or not. Flickering clusters of cellular water molecules set up a synchronized rhythm with which all structural molecules danced in harmony. Once a cell was dead, its organization fell apart. Although the clusters of water molecules continued to vibrate, their influence was no longer ordered and all the disintegrating cell parts bobbed about at random. To this degree, water in living cells, derived from lake, stream, and rain, was life itself.

The demise and disintegration of a cell, in the presence of decay bacteria living in water, was no catastrophe but a promise of future life in some other form. The dead cell's compounds of nitrogen, phosphorus, and other elements set free were soon caught up in further chemical and biological events which made it possible for organisms to build, grow, work, and reproduce. The same ingredients had been used and reused for over three billion years.

During its years of isolation the pond had changed. No longer were its plants and animals the same as those of the stream. When it had been cut off from the meandering flow, imprisoned stream animals disappeared as the stilled water failed to bring the food or provide the immense quantities of dissolved oxygen they required. Soon the new

pond attracted other forms of life to an existence not available elsewhere on the valley floor. Aquatic insects flying
overhead plummeted toward the bright reflective surface
until the water was heavily populated by whirligig beetles,
back swimmers, water boatmen, and diving beetles. On the
legs of some insects came the larvae of red water mites and
crustaceans, but examples of transfers of this sort were
commonplace. Each heron wading into the water, each
duck that came skittering in, brought eggs, spores, seeds,
whole organisms, and viable fragments of living things on
their feet and feathers. An occasional killdeer stepping
lightly into the shallows might leave behind a snail which,
because it was both male and female, could give rise to
unlimited progeny. After even a short time of this kind of
immigration, life in the pond was intense. Seeds and spores
had grown into plants that rose from the mud; small animals crawled over submerged vegetation and upon the bottom; clouds of plankton drifted through the water; predatory fishes entered by eggs, or through rivulets between
stream and pond, and had survived.

The pond was not an entirely isolated microcosm, but
was involved in a continuing exchange with the world
around it. Every rain brought contributions from the land
—minerals, silt, debris, and life. The shoreline was blanketed by old and decaying leaves that had blown down,
become saturated, and slipped beneath the surface the previous autumn. Insects flew in to lay eggs that would hatch
and grow as aquatic larvae, to leave later, taking with them
the product of a season of nourishment and growth. No
sooner did most insects emerge from the water than they
were snatched in mid-flight by kingbirds and swallows
swooping over the lake in daylight, or by low-flying bats
after dark. The pond was a major contributor to the valley
world as well as a recipient, providing food for ringneck
ducks as they dived to take molluscs and insects off the
bottom, and for mergansers that caught fishes in deeper

water. Surface-feeding mallards and teal came to consume vegetation and raise their young in the shallows. Occasional ospreys snatched large fishes basking at the surface, and the many herons present speared fish, frogs, and even water snakes that had come to hunt crayfish. Raccoons walked the shore every night to scrabble for clams and crustaceans.

A lake or pond has an existence that simulates life itself. It is born abruptly, matures slowly, and eventually declines into senescence. Finally it dies, its grave only a richer shadow upon the valley floor where once it gleamed in the sun.

The basin of each very young lake in the valley held open water with little sediment lying on the bottom. At times the entire valley floor had been inundated by surging floods, when fine particles of suspended clay settled into the crescent-shaped depressions, carpeting their basins with a thick ooze. Several of the ponds possessed small tributary streams that carried additional loose material into the tranquil water. The oldest ponds lay quiet and thick with a suffocating blanket of clay and organic soil, a deposit often reaching within inches of the surface; it was here that dense stands of aquatic vegetation grew high above the shallow water. The luxuriant growth of these plants, filling old ponds, hurried them toward becoming swamps and bogs, and inevitable extinction.

Up and down the valley the outlines of old meanders and ox-bow lakes were defined by parallel rows of vegetation following the contours of diminished shorelines. The open pond in this one pocket of the valley was neither new nor old, but mature and healthy, supporting a multitude of lives.

Grasses and low shrubs carpeted the valley; thickets of greenbrier looped between the bases of young trees just gaining a foothold in the rich soil. Where the ground was moist near the pond, the rich rose-purple flower clusters of swamp milkweed provided a background for the vivid accent of golden spatterdock. Tall willows had taken root along the shoreline where a few other moisture-loving trees provided shade and shelter. Blue iris and royal fern, the principal fern in the low valley, grew side by side in small clearings along the banks. A solitary muskrat, glossy fur not yet reflecting the sun just appearing over the low hills, pulled down leaves of sweet flag, savoring their sugary contents. In an open spot, twelve-foot stalks of reed grass, topped by feathered tassels, bowed slightly to the gentle morning breeze. A long-billed marsh wren clutched one tasseled stalk, bobbing back and forth, before darting to a nearby buttonbush at the waterline where the tiniest of ripples lapped the narrow silty beach. The buttonbush displayed spherical clusters of flowers, which later would turn into fruit sought by swamp sparrows and by muskrats. Solitary erect stems of marsh horsetail rose straight into the air, hardly changed from their ancestors that had lived long before there were seed-bearing leafy plants, many millions of years ago.

Where the water was only an inch deep, dense stands of spike rush bore tiny fruits, many now damaged and removed, taken by waterfowl, marsh birds, and the ever-hungry muskrat, who depended more upon aquatic vegetation than any other mammal of the northern continent. In shallow water along the shore, arrowhead and pickerel weed emerged to spread their broad, spearlike leaves a foot or more above the surface. From the pickerel weed vertical clusters of blue flowers arose, but were dwarfed by the towering flame-red spikes of loosestrife behind them.

The two plants of the shoreline covering more territory than all others were cattails and waterwillow. Long flattened swords of cattails fanned out from submerged, starchy rootstocks; some of the leaves extended over six feet into the air, providing shelter for a bittern which made its quiet and deliberate way through the dense vegetation in search of prey. The bittern paused briefly beside a vacated nest of a black duck crowning an old tussock; there it found a recently emerged leopard frog, still with stump of a tail from its tadpole days. The long beak of the bird flicked out, removing the frog before it had a chance to react. With head low and outthrust, the bittern stalked on.

Waterwillow, with its tough, woody, looping stems, was one of the major agents in the eventual demise of the pond. With each year's growth its stems extended over the water, took root, until there was a thicket of almost impenetrable vegetation along the shore. Wherever stands of waterwillow were established, sediment accumulated to the surface and above, diminishing the shoreline year after year.

In a few places, waterwillow was covered by a tangled orange mass of vines, the webbing of dodder, a parasitic plant. Although dodder produced small and attractive white flowers, it had lost chlorophyll and all ability to manufacture its own food. Instead, it grew protruberances that tapped the food-conducting vessels of waterwillow stems, a condition tolerated by the host plant.

It was the plants of the shoreline that gave this and every pond a character of its own, and established suitable conditions for a multitude of inhabitants. Depending upon what vegetation grew and how abundantly, waves were reduced, light filtered, and refuge from predators was made available to lesser creatures of the watery world. Some animals used plant materials for nests and cases into which they could retreat, while others, attached to plant stems, could be transported along the shore if the stems broke free and drifted away. Rooted plants consumed nutrients from soil and water, freed oxygen into the pond and the air above, consumed quantities of carbon dioxide in daylight hours, and left their remains to enrich the shoreline. These larger plants also helped keep the water clear, and not only by minimizing wave disturbance of the muddy bottom: they gave off substances that reduced the surface tension of the film, allowing particles to drift to the bottom more quickly.

Because of the exceptionally high food content of many
plants growing in and near the water, the pond drew vegetarians from all over the valley floor, far more than could
be supported by the relatively sparse plant growth along
the shallows of the rushing stream. Some of the softer
plants, well out in the water, were eaten regularly by
dozens of different animals, especially insects and turtles. A
few fishes were vegetarians, as were several kinds of water
fowl that were frequent visitors. Other than the ubiquitous
muskrat, however, few mammals came to the pond to feed
—only deer and an occasional venturesome rabbit. Where
beavers were established in the watershed, they too consumed vegetation, but they had not traveled widely and
were absent from this ox-bow pond.

Insects were abundant within the zone of emerging
aquatic plants. Now that the sun had risen and the fog
lifted, the shoreline was busy with darting, brilliant-hued
damselflies and the flicking, powerful flight of dragonflies.
A pair of damselflies hung over a cluster of sedges, the male
grasping the female's head by a pair of pincers at the tip of
his abdomen. They alighted together on a single stem, the
female immediately backing downward into the water until
she was completely submerged. There she deftly deposited
eggs on the plant, but without the male would have been
trapped beneath the tough and elastic surface film. He
crawled and fluttered upward, still clinging to her, until she
broke free and they flew off to repeat the process on
another plant.

Most of the dragonflies flew in great, looping patterns in
pursuit of air-borne prey, often coming to rest on a bare
shoreline stem; there they crouched, outstretched wings
glinting in the sunlight, while their head with huge bulbous
eyes swiveled rapidly, following the flight of a possible
insect victim. Other dragonflies were occupied with a more
regular flight, as they had been in the nearby stream, flying
back and forth along the shore, facing the water and de-

fending their established mating territories against invaders of their own kind.

Beneath the surface, predatory diving beetles and back swimmers darted about in search of smaller prey, although one very large beetle had just seized a tadpole twice its own size and was tearing it to pieces with its powerful jaws. Water scorpions, seemingly lethargic, hung close to the surface with raptorial claws cocked back in readiness. Water boatmen, insects bearing a superficial resemblance to back swimmers with their long sculling legs, swam to the bottom where they swept algae and microscopic animals from the fine sediment. At times a water boatman fed directly upon algal filaments through a piercing, protrusible tube that permitted it to suck out plant juices.

Many submerged insects carried air with them as silvery bubbles or as shining coats covering their bodies. This air was not consumed directly, but served as a unique, nonliving gill. The surface film of each bubble was a boundary membrane between the aquatic world and the dry atmosphere of the insect's breathing apparatus. Oxygen, dissolved in the pond, diffused readily through the film into the bubble where it was drawn upon for respiration. Carbon dioxide, in turn, was diffused outward as it accumulated as waste. Some insects carrying such physical gills beneath the water could remain submerged for hours; indeed, a few small kinds never replenished the silvery coatings their entire adult lives.

The success of other swimming insects in obtaining air was due to long tubular snorkels fringed with water-repellent hairs around the aperture. When the snorkel protruded through the surface film, not only was water prevented from flooding in, but the hairs functioned as an anchor, keeping the animal locked in place while replenishing its air.

Fisher spiders shared with their distant insect cousins the ability to carry air beneath, although they did so only rarely and then when pursued by larger predators or when they in turn had an opportunity to capture a small fish. Otherwise they rested upon the surface, all eight legs radiating flat across the water, depressing the film slightly with their weight.

Through the corridors of quiet water formed by the erect blades of cattails and arching waterwillow stems, whirligig beetles coursed busily and water striders skated, resting for long intervals between each sudden thrust of their legs. Beneath them gray bullhead catfish grubbed among plant fragments littering the bottom, oblivious to the insect shadows on the mirrored ceiling directly overhead. Two satinfin shiners swam slowly by, gulping plankton that swarmed in the shallow water; they served as important links between the microscopic world and larger predators that sooner or later would catch them. A lone mud pickerel, only inches long, hung obliquely in the water, head up, waiting for an insect to come within range of its sudden, inescapable rush.

The most common fishes in the warm, lighted water were yellow perch and sunfishes, especially crappies, bluegills, and a few brilliant blue and orange pumpkinseeds with scarlet tabs on their gill covers. Sunfishes, characteristic of lakes and ponds, lacked the streamlining of a brook trout with its strong lateral muscles necessary for living in swift water. Their bodies were compressed and deep, allowing them to slip with ease through the tangled masses of aquatic plant stems.

A pair of pumpkinseeds had just finished clearing a rounded depression in the muddy bottom close to shore, creating a light spot where bared sand and pebbles contrasted with the dark organic debris everywhere else. The female was passively circling the nest, the male hovering on her right side, slightly to the rear. As he drew even closer to her side, he tilted away from the vertical, slanting under her belly, ready to release sperm-containing milt whenever eggs were produced. This action occurred repeatedly, occasionally interrupted by an irritated rush from the female, driving him away a short distance. He immediately returned and persistently began the process over again. Before long, the female laid her first eggs, cementing them to clean pebbles on the bottom as the water grew cloudy with milt from the male. As soon as she finished, she took up a position in the center of the nest, fanning the bottom with her fins to keep silt from settling and to provide a steady flow of oxygenated water to the eggs, in which each original cell was already beginning to cleave into two. The male would spell her from time to time, in addition to being an aggressive protector of the nest, charging all intruders and driving away most. But when one of the parents left the area for a while, the other was not always successful in defending the brood. A minnow dashed in, scooped up an egg, and fled, pursued briefly by the remaining sunfish. Even with the nest left unattended so briefly there was sufficient time for other minnows, waiting nearby, to dash in and gulp as many eggs as they could.

Small leeches clung to the tails of several sunfishes, streaming out like banners from the fan-shaped fins. Other leeches looped along the bottom, searching for food in the mud. A large horse leech swam rapidly past in mid-water, its long body waving up and down like a flat ribbon, displaying a bright orange belly.

The underwater portions of all emergent plants were heavily populated. Life on submerged surfaces of stems, old sticks, rocks, and debris composed a distinct assemblage, quite unlike all other zones of life in the small lake. Plants and animals living there had to be fast-growing, as their restricted world lasted for little more than the warm months of summer. Some were firmly attached, but many others were not; all depended upon light, moderate temperature, and the slight water movements of the shallow pond. When a fish or turtle swam violently by, some of the encrusting plants and animals were swept off and became members of the drifting life of the plankton. Others retreated into tubes, sheaths, or cups, or were held firmly attached by tough stalks and soon returned to full feeding activity after being disturbed by strong local currents.

Dark smears of blue-green algae wrapped some of the stems, but on most a loose brownish coating was composed of microscopic diatoms, tiny plants housed in glasslike cases of geometric precision; like those of brook and stream, they were fresh-water relatives of the diatoms that had drifted under the surface of the sea where the river had begun. A host of one-celled animals, hydras, insect larvae, and worms burrowed into the algal crusts, seeking food and protection. Small crustaceans scrabbled across the coated stems, interrupted at times by the slow coursing of larger snails rasping away the film of diatoms with a flexible, toothed tongue. Clear, elongated masses of snail eggs were everywhere; inside the transparent egg spheres tiny embryonic snails rotated smoothly and ceaselessly. Long green dragonfly nymphs, the largest predators on the stalks, waited to shoot out hinged jaws until smaller, unwary victims came within range.

In some places the stems were heavily encrusted solely with animals, fixed permanently in place, at least for the duration of the season. Fresh-water sponges created a rough and irregular mass, hugging the stem without rising in fingerlike projections as had the stream varieties. Other stems were enveloped by great gelatinous chunks decorated with hundreds of dark rosettes. Each rosette was composed of several soft, tube-dwelling moss animals that extended tentacles arranged in a half-circle; tiny beating hairs on the tentacles produced currents which flowed toward a centrally placed mouth, where quantities of minute plant and animal cells were removed from the water.

Short mud-coated tubes lay along many of the stems. Inside each was the flickering, waving body of an elon-

gated midge larva. Sinuous motion was its means of drawing a current into one end of the tube, bringing with it
oxygen and food.

Although the shoreline of the ox-bow followed a smooth
arc, the vegetation growing out into open water did not.
There were places where waterwillow was indented toward the shore, forming coves in the tangle of plants. In
such protected spots the surface was covered by a blanket
of tiny floating plants, duckweed and water meal, the
smallest flowering plants in the world. The grain-sized
water meal had no roots at all, but duckweed, a trifle larger,
had fine, unbranched rootlets hanging a quarter of an inch
below the buoyant leaves. The dense unbroken blanket
enclosed a vast population of smaller plants and animals, of
which some attached to the tiny roots while others swam
safely through the maze of miniature vegetation. Insect
eggs and larvae, worms, tentacled hydras, snails, one-celled
animals, round-bodied water mites, crustaceans, rotifers,
algal filaments, and single-celled diatoms and desmids lived
together in profusion and endless variety. But the pond
bottom only a foot below was not so heavily populated, for
sunlight failed to penetrate and little oxygen was present on
the muddy floor.

The blanket supported another population on its upper
surface, mostly insects and spiders that ran across the dry,
cobbled surface of the tiny plants. Flies, mosquitoes, and
damselflies alighted, depressing the matted surface slightly.
Far more abundant were the springtails and jumping plant
lice, extremely small insects that lived their entire lives in
this environment. Disturbed by a turtle thrusting its head
through the blanket, they bounded into the air in clouds of
light specks describing arcs in the early morning sunlight.
The turtle, head coated with duckweed, blinked a few
times attempting to clear its eyes, but soon submerged,
leaving no trace of its interruption as the floating plants
spread once more across the surface film.

Heavy masses of filamentous algae, rather than duck-weed, lay in a few of the coves created by indentation of shoreline plants. Glistening hemispheres of gas, greenish from the fine threads that kept them trapped, extended above the surface. Underwater the scene was one of deli-cate beauty, for the filaments rose gracefully from the bottom in long streamers, usually with an apex formed by a single luminescent bubble held captive in its attempted ballooning ascent. The algal veils were almost still, but swayed slightly in the vague currents of the pond. Several shiners appeared and vanished behind the green curtains, their silvery sides glinting momentarily as they caught the sunlight. The filaments were clean of attaching organisms, unlike the roots of duckweed, although a few specialized creatures moved along them, feeding on the tubular cells by piercing their walls and sucking out the contents. One of the most highly adapted of these was a rotifer which wrapped two lobes loosely around an algal strand and swam briskly, guided along by its living track.

Beyond the cattails and waterwillow, the character of shoreline vegetation changed. Where the pond bottom sloped and grew deeper, plants capable of holding their foliage above water could no longer exist. Instead, plants rooted in the bottom sent up long flexible air-filled stems that terminated in floating leaves. Pond lilies and smart-weeds were the most common, with a few pondweeds growing near the outer bend of the ox-bow. Both the smartweeds and pondweeds produced two kinds of leaves: long, slender ones underwater that offered little resistance to currents and passing animals, and broad, flat leaves of great buoyancy floating on the surface. The cordlike stalks and stems of these plants, as well as the thicker stalks of pond lilies, were tough enough to stand almost any amount of wave action, and were coated with a mucus that reduced damaging abrasion when they rubbed against one another. Because the buoyant stems of the larger pond lilies contained many air spaces, certain aquatic insects deposited their eggs deep within, insuring a protected, oxygen-rich environment for their vegetarian young.

The dark water of the pond was spotted with the white flowers of water lilies, flowers which had just opened in the morning light to remain wide until midafternoon when the petals would close tight again. The plants from which the lilies grew lay well beneath the muddy pond bottom, horizontal stems heavy with starch deposits, equipped with short stubby roots that groped even farther into the soft sediment.

A remarkable feature of all floating leaf plants was the design of the leaf itself, with its air cavities that assured buoyancy. The bottom surface of each leaf was wettable, but the top was not, repelling water with a waxy coating. The leaf therefore was always exposed to light and air; even if turned over by a gust of wind, it soon righted itself by its properties of attraction and repulsion of water. Although the leaves derived additional support from the strength of the surface film, they imposed a penalty upon themselves by releasing substances into the water that reduced the surface tension slightly.

Underneath each lily pad was an assemblage of creatures and plants resembling that of the duckweed community, but because of the size of these leaves, much larger forms of life were included. Dozens of translucent hydras hung down, trailing long tentacles into the murky water beneath. On most of the hydras, small spool-shaped one-celled animals scurried about, finding protection within the range of their host's stinging tentacles and an opportunity for capturing food. Sponges, moss animals, fresh-water limpets, and a host of insects competed for space in this upside-down world. Some of the larvae mined the lily pads, leaving long sinuous cuts through which sunlight stabbed into the darkness below. Other holes had been cut by beetles alighting on the upper side. One such beetle had just completed a smooth round hole and now inserted her abdomen in it, depositing a semicircle of eggs on the undersurface where they would later hatch into aquatic young.

The small minnows of the cattail and waterwillow zone were no longer common in deeper water, but sunfishes still threaded their way among the curving stems, keeping close to the surface, while predatory largemouth bass hunted near the bottom. Midway a large chain pickerel hung almost motionless, eyes directed forward along two grooves down its snout, focused on a young bluegill beginning to come in range. The sunfish veered away, but the pickerel did not give chase, remaining motionless except for a gentle fanning of its fins.

Floating leaves were only one indication of the support offered by the surface film, the result of strong attraction of water molecules for one another. All molecules directly at the surface were linked to their lower neighbors through the hydrogen they carried, four bonds to each water molecule. The effect was a strong and elastic film that supported distinct populations not only on top, but directly underneath, serving either as floor or ceiling for a multitude of living things.

The surface of the ox-bow lake was of vital importance to its physical character as well as to its inhabitants. Its great expanse allowed a continuous passage of gas molecules back and forth between water and atmosphere. It was important to aquatic plants that water be a capacious absorber of carbon dioxide. Aquatic animals, on the other hand, could not tolerate too little dissolved oxygen, a gas available only from the activity of plants. Some oxygen came from algae of the pond, but most filtered down into the water from the atmosphere, where it had traveled from its origin in forests and fields or in the microscopic plant life of a distant ocean. When wind ruffled the surface of the pond, gas exchange was enhanced, but never to the degree of the turbulent mountain brook, or even of a riffle in the nearby stream.

The morning air was quiet and the water surface yet undisturbed. When the sun had risen a few hours earlier, most of its light had been reflected off the still water, but as the sun climbed higher, increasing amounts of light entered the pond, were deflected through refraction, and sank downward. With so many rooted plants, the water was quite clear and the dark bottom now was illuminated with a subdued golden glow. The first stirrings of a gentle morning breeze came down from the hills, rippling the water and creating bright dancing lines upon the muddy bottom.

There were seasonal effects at the surface: light fell at different angles in winter and summer, and with lowered temperatures, molecules of the water grew less active and more densely packed. When the chill winds of winter cooled the surface even more, the increased weight of the upper layers would cause water to sink, creating an overturn that would bring warmer water up to be cooled in turn. In deep lakes which were distinctly layered all summer, the overturn was a phenomenon of major importance, but the ox-bow lake was shallow and more of a pond than a lake, so temperature throughout its slight depth soon followed that of the atmosphere.

Once the pond was thoroughly chilled and temperatures continued to fall in the air above, molecules of the surface would arrange themselves into loosely packed crystals with air spaces in between. Trapped air meant the ice would float, rather than sink as most solids, and would nearly seal the pond from further atmospheric influences. Gas exchange would diminish, seriously affecting life in the pond, perhaps even suffocating some of the fishes that required more oxygen than the water contained from an earlier season.

On top of the surface film, the smallest products of the

plant world coated the water with a dusting of pollen grains and spores. They were important as food to a few of the creatures living beneath that were able to harvest them.

Creatures of the surface lived not on wet water but upon a smooth, yielding film, prevented from plunging through by water-repellent pads on their feet. The normal animal inhabitants were mostly fisher spiders and insects—whirligig beetles with their divided eyes, one pair up and another pair directed down into the water, water striders, springtails, and water measurers. But there were also accidental members of this population in the presence of crickets, bees, and land beetles struggling as captives of the film; only a few were finally able to climb upon a leaf or floating twig and so escape the clinging grasp of the surface. An elongated rove beetle fell from an overhanging branch, landed on the water, and immediately shot across the surface, quickly coming into contact with a stem of water-willow. It had no apparent means of propulsion, but emitted a wetting agent that destroyed the surface tension in front of its body; the pressure from the remaining tension against the rest of its body pushed it forward without any particular accuracy, but considerable speed.

On the underside of the film, plants were mostly single-celled algae and bacteria, but animals lived there in great variety, hanging down or crawling across the mirrored ceiling, only slightly dimpling it with their weight. Pond snails and flatworms flowed evenly across the wide expanse, the snails every now and then dropping straight down to the bottom several feet below. Other snails, having generated a gas bubble beneath their shells, rose slowly in stately fashion to the surface. A few very bright green hydras hung from the film, their color due to microscopic algae living within their cells. Specially adapted water fleas held to the surface, upside down, with bristles caught in its tension; they rowed along rapidly with beating antennae, scooping pollen grains from the other side.

For other water fleas living as members of the plankton lower down, the film was a great hazard when they found themselves accidentally thrust above it. If they were trapped in its firm grip, there was only one means of escape: they might molt their outer shell-like skeletons and slip down into the fluid water beneath, leaving the tiny transparent empty skins on the surface glistening in the sunlight. The cast skins, along with pollen, fungi, and other organic bits floating on top of the water, were collected indiscriminately by mosquito larvae cruising just beneath the film. The larvae were caught in turn by minnows and predatory insects rising from below.

The last zone of vegetation extending into the pond was composed of deep plants, rooted in the bottom yet never reaching the surface. Most were slender and simple, but one, the bladderwort, was one of the most unusual plants in the whole ox-bow pond. Underwater it appeared delicate and vinelike, bearing many rounded bladders, each with a mouthlike opening guarded by sharp, bristling hairs. As the plant grew, its bladders were first of all small and green, turning red with age and an increase in size, finally becoming almost black. Its presence was a sure indication that plankton was abundant, although in the dense tangle of bladderwort that filled some shallows of the pond, the traps were so efficient that the plankton population was noticeably diminished in the immediate vicinity. Whenever a rotifer or small crustacean entered the slit of a bladder, its escape was blocked by the hairs and eventually it succumbed to digestive juices of the plant, although perhaps not for a day or two. In some bladderworts, capture was made more certain by a sudden inflation of the bladder, sucking in water and the nearby animals caught in the flow. That the traps were effective was attested by the presence of ten or a dozen small creatures in most bladders, the largest of which could capture this number in less than two hours.

Why did such wholesale trapping take place? Animal food was not absolutely essential to bladderwort, but if one of the plants was denied sufficient quantities of plankton, its growth was stunted; it flourished where microscopic life was abundant.

Because of a general lack of currents in the small lake, clouds of plankton drifted through the open water away from shore. These, the smallest lives in the ox-bow, were present in every possible shape and degree of specialization. Close to the surface, green spindle-shaped cells swam, each propelling itself with a single hairlike organ, although another tinier flagellum extended partly from the cell body. The greenness was that of chlorophyll, showing a dependence upon the sun in the manufacture of food from water and carbon dioxide, but the motile whip suggested an animal trait. In the front of each of these cells was a red eyespot, slightly off center; when the creature rotated on its own axis, the eyespot was directed now at the lighted surface, now at the shadowed bottom, causing corrective and appropriate action to be taken by the flagellum to insure the cell's remaining at the surface.

There were many other such cells, but not all green and not all solitary. Some lived in little brown bottles of their own manufacture; others clustered in spherical colonies in which there was some degree of coordination to allow the entire sphere to behave as one.

Plant plankton, whether single cells or many, motile or completely passive, were the dominant forms of drifting life, not only because they were the most abundant. By manufacturing food from simple raw materials in the water, they formed the base of all food relationships in the pond, no matter how remote or obscure. They were essential links between radiant energy of the sun, and the multitude of lives this one small body of fresh water supported. They were grazed upon by planktonic vegetarians, which in turn were caught by slightly larger predators, continuing through a series of events to the largest of the pond's carnivores—fishes, turtles, frogs, and birds. But at every step there was a tenfold loss of available food energy, so the large predators were few, while the tiny plants blanketed the pond in drifting pastures of incalculable numbers.

Rotifers, common animals of the plankton, were not much larger than the plant cells, but were composed of several dozen cells, the number always the same for each different kind. A rotifer swam vigorously by extending two rounded crowns out in front of its head, each fringed with short hairs beating in synchronized waves. Inside the transparent body of a rotifer was a deeply seated pair of complex jaws, busily grinding fine all food particles swept into the mouth between the two crowns.

Tiny crustaceans were even more abundant and were more complex despite their small size. Most swam by means of Y-shaped legs or muscular antennae which rowed or beat the water, thrusting the animals along in a jerky fashion. The smallest crustaceans had finely feathered legs beneath a teardrop-shaped body; legs with such closely spaced bristles that individual bacteria were filtered out of the water to serve as the prime constituent of their food. Water fleas, also small crustaceans, were oval and compressed, each with a short projecting beak, above which a single multifaceted eye quivered and rotated in its internal socket. A few water fleas carried a saddlelike chamber giving them a humpbacked appearance; within the chamber were two or three eggs or developing larvae, already beginning to kick and move.

The variety of other animal plankton was almost endless. In addition to the more common crustaceans, one-celled trumpet-shaped creatures swam smoothly along, most of them a delicate blue; there were scarlet water mites whose spherical bodies tumbled through the water; larval parasites seeking a host; tiny immature insects; and juvenile fishes, their bellies still swollen with yolk. The numbers of each kind of plankton organism waxed and waned with seasonal events; what was common today might have been rare a week before, and almost certainly would be diminished a month hence.

The distribution of most of the plant plankton, and many of the animals, was influenced by temperature. In the cold, dense water of fall and winter, the tiny organisms were more buoyant than they were on this midsummer day when the water was warm. If they lacked special adaptations and were unable to swim, they tended to sink. Some of the small crustaceans would swim obliquely upward, then drift slowly down to a certain level determined either by pressure, coolness, or a lessening of light before they became active again. Other plant and animal plankton were able to lessen their weight and approximate the density of the water by manufacturing oil droplets that served as minute buoyant balloons, encased in their bodies. Most of the geometric plant cells, diatoms and desmids, not only had such droplets, but extremely thin shells to diminish their weight. Animals eating these plants often incorporated the oils directly into their own bodies, so they too were not as heavy as they seemed. A few planktonic animals manufactured gas bubbles which kept them afloat even more effectively. Other members of the plankton community relied upon increased friction to resist sinking: with long hairs, flat blades, elongated bodies and appendages, body surface area was increased until they sank far more slowly than organisms with a more compact shape.

The gentle eddies created by wind and temperature differences in the pond aided in keeping plankton suspended, but did not cause enough mixing to prevent a certain amount of layering among the population. Some microscopic plants were always close to the surface, but others flourished only in deeper water just off the bottom, where the water was a little cooler. Animal plankton, with their capacity for independent movement, migrated vertically each day, rising at night and sinking during the daylight hours, each form leveling off at the layer most favorable to its particular needs and its stage of development.

Throughout the curve of the ox-bow the plankton populations were not evenly distributed either in depth or in area. In certain places, available nutrients had been depleted by microscopic plants. As plant numbers fell, a decline in the grazing animal plankton followed. In other spots, nutrients continued to be abundant, perhaps flowing off the surrounding land, and the drifting populations flourished. No matter what the conditions, their numbers and variety far exceeded anything the parent stream had so far produced, or would produce until meeting the sea in a distant bay.

One insect larva, that of the phantom midge, linked both the surface plankton and the murky bottom in which it burrowed out of sight during the day. At night it rose close to the surface where with quick, shimmering movements it thrust its transparent body in pursuit of smaller, weaker animals, which it captured with a great hooked beak. At either end of its body, the larva contained a pair of silvery air sacs. The pressure within the buoyant sacs was altered to keep the insect in one plane, so its energies were expended for rapid horizontal movement rather than overcoming a gentle gravitational pull.

The thick sediment where the midge larva took refuge during the day was alive with organisms, especially near the boundary between mud and water. Thin, smooth roundworms squirmed among the fine particles; water bears crawled through the ooze on stumpy, clawed legs. Bloodworm midge larvae vibrated in loosely constructed mud burrows, using a hemoglobinlike pigment to extract oxygen from the impoverished layer of water lying above the muddy silt where this vital gas was exhausted by bacterial decomposers. Decay was the single most important occurrence on the bottom, for it allowed the recirculation of essential compounds into the world of life.

Segmented worms burrowed head downward in softer sediments, extending the upper half of their bodies out of chimneylike tubes that rose above the mud surface. Here they performed an important function by turning over the bottom sediments; without such activity, the bottom would build up more rapidly, hastening the demise of the pond. When the oxygen levels were especially low, these tubifex worms undulated briskly, creating currents of water that helped circulate what little dissolved gas was available, until the bottom appeared to flicker with the pale flame of their yellowish bodies. There were other curiously shaped segmented worms: some with a long, drawn-out proboscis that nervously probed the sediment; others with tufts of gills at the end of their transparent bodies; some predatory, and some that fed upon organic matter of the bottom much like their larger earthworm relatives on shore.

Flattened dragonfly nymphs walked slowly, spraddle-legged, to keep from sinking in the loose sediment. Each breathed by drawing water into a capacious chamber in the rear of its body. Usually the insect expelled the water gently but, if danger threatened, it could be shot out with enough force to propel the nymph out of harm's way. Crustaceans of several kinds occupied the same soft bottom layer as the nymphs, sifting through the organic ooze for nourishing particles. Some were small with two hinged shells like a miniature clam; others were larger and lay on their sides, kicking vigorously until clouds of sediment rose about them.

The nature of the bottom varied with depth; in the shallower approaches to the shore, with oxygen more plentiful, animals increased in numbers and variety. Amoebas and their shelled relatives groped through the fine silt, rounded pill clams strained the tiniest suspended plankton

through their long, pink siphons, and loose particles resting
on the mud surface were worked over by a multitude of
crustaceans, large and small.

Over long years of settling, the finest sediment lay in the
center and deeper parts of the pond, but gravel, sand, and
lime marl produced by plants grew more pronounced
closer to the shore. Plant fragments and fallen leaves, and
all kinds of drift material, carpeted the shallow bottom
along the shoreline. Life was abundant here: plants manufactured food from dissolved gases and minerals in the
lighted zones, while animals depended mostly upon the fine
decaying organic litter for nourishment. The wandering
furrows of snails crisscrossed the fine sediment, sometimes
ending abruptly as a snail released its hold and slowly
ascended to the surface, buoyed by a bubble of trapped gas.
At the surface, it performed a leisurely somersault, ex-
changed the waste gas for a gulp of fresh air and sank
gently to the bottom, where it commenced feeding again.

The bottom of the ox-bow was in fact a region of intense
life activity, where the dominant organisms were bacteria
and one-celled animals. In deep places so little oxygen was
available because of a lack of circulation in the water that
decay was unable to proceed all the way, resulting in odorous gases that at times rose as bubbles to burst audibly at
the surface.

Bacteria were everywhere in the pond, giving many of
its zones their distinguishing features. As members of the
plankton, they were common near the surface, not so numerous in mid-water, and extremely abundant on the bottom. No matter where they lived, they were active in the
decay of plants and animals, in the conversion of organic
substances to dissolved nutrients and gases, in the trapping
of nitrogen into usable compounds, and as parasites in
other organisms.

The shoreline with its plants, the surface film, and the
bottom with its burrowers and bacteria were only the mar-

gins of the pond; its bulk lay in the water itself. The most numerous creatures of the open water were plankton, but they were small and indistinct, clouding and coloring the water a pale yellow-green. The large and more visible animals of the pond were far fewer in number, often drawing near to the plant-choked shore for food and shelter. There were wanderers among them, especially the predator bass and pickerel that followed the curved bed of the ox-bow from one end to the other. Turtles swam freely, popping up for breath anywhere across the still surface. Now that the sun was high and gave great warmth, every log and protruding rock or stump had its full quota of basking turtles, some lying on top of others in their attempt to absorb heat. Small, dull mud turtles and musk turtles lay close to the water; but higher up, often balanced precariously, were painted turtles and even great red-bellied turtles, stretching their legs into the air, allowing the sun's heat to penetrate their blood streams where it would be transmitted to the rest of their heavy bodies. Without an ability to regulate temperature, turtles had to rely upon the heat of the environment to maintain a sufficiently high pitch of bodily activity.

With slow and majestic wingbeats, a great blue heron cruised overhead, flying low over one crowded log, provoking the turtles into action. The red-bellies were the first to go, sliding into the water with great splashes that awoke the others to possible danger. Within a few seconds all had slid into the water, sculling deep for safety. But the bird pumped on overhead, and soon one turtle head after the other broke through the surface to inspect the air and shore. Slowly, laboriously, they clambered out on the log again, one often upsetting the balance of another, sending it careening into the water before it had a chance to dry off. Within five minutes, all were back drying, warming, stretching necks and limbs to the bright sun.

This was the pond, unique in the world, yet possessing

patterns repeated a million times over in the northern continent, patterns that could be seen in all the ox-bows of the
river valley. Like the stream, it was a world of water, yet
different in the opportunities offered to life. It had originated with the stream and might die without returning to
it, but in more than one place the stream had changed
course and abruptly entered an ox-bow it had forsaken
many years before. At the lower end of the pond, a meander was inching back into the dike it had built; now it was
separated by only a dozen yards.

The proximity of pond to stream allowed some animals
and seeds to be transported from the still water across the
narrow wall and carried far downstream, perhaps lost forever to a successful existence, but also possible colonizers in
another ox-bow about to be cut off miles away.

The stream passed on beyond this quiet pocket of water
and the life it supported, beyond other ox-bows declining
into senescence and a few recently created. Like a great,
restless thing, the stream snaked through the years back and
forth across the valley, breaking off fragments of itself that
would exist for brief moments in time and then vanish.
These children of the stream were the only means by
which water, in its headlong rush to the sea, could rest
upon the land.

VII

TO JOIN THE BRIMMING RIVER

. . . the current, an emblem of all progress, following the same law within the system, with time, and all that is made. . . .

HENRY DAVID THOREAU

T H E huge presence of the river was a dominant feature of the lowlands, drawing from the bordering hills contributions of lesser rivers that joined in one great surge toward the coastal plain. It seemed timeless, but the present river was not very old geologically, and had been preceded by others in former ages. When the last ice front retreated fifteen thousand years ago, it left a cold lake, one hundred miles long, held behind a dam of rock and accumulated drift materials. Originally the lake had been over a hundred feet deep, but during a period of four thousand years sediment built up from the bottom, alternately depositing light-colored silt in the summers and thinner layers of dark clay in the winters. Suddenly, during a spring thaw long ago, the entire drift dam collapsed and the lake broke free, rushing down the valley that soon was drained except for a central channel.

By the time the lake disappeared, the glaciers had withdrawn far to the north, leaving scoured valleys in their wake, each with its river draining ice water from the diminishing ice cap. When the glaciers were still close, early plant colonizers of bordering slopes had been poplar, cranberry, and arctic willow, but before long they were replaced by birch, spruce, and pine, which spread across the newly exposed valley floor. Later, as the climate moderated, alders, hazelnuts, and elms appeared, followed finally by oaks, which now were abundant. The present river forest was rich in diversity and thick with vines shading a dense undergrowth of moisture-seeking violets and other small ground plants.

As it grew warmer following the glacial epoch, the river came to support a profusion of aquatic plants in shallow water, some of which were still preserved intact after thousands of years in clay banks. These near-fossils differed not at all from their modern descendants that now grew in the warm, illuminated water close to the shoreline.

The history of the river since the retreat of the glaciers was not a simple one, for its course had changed many times. Once it and its future tributaries had flowed parallel to one another, but some of the streams had been diverted to flow at right angles by tilted dikes of hard rock alternating with softer layers. When one river cut across to join another, the captured stream was forced to follow the path of the pirate river. The valley cut by the ousted stream over thousands of years was now bereft of water and became a dry, windy pass sought by migrating land animals and low-flying birds. Several gaps in the hills adjacent to the great river gave mute testimony to its past conquests.

For millennia the river had cut back its valley toward the headwaters, becoming more gradual in its descent, widening and deepening its channel. The longer this occurred, the fewer the riffles that broke the surface, and the river grew placid in its might. The prolonged development of the river, proceeding from its mouth toward an upland source, was defined by plants and animals slowly invading regions formerly too turbulent for their kind.

The river, having drained seven thousand square miles, now flowed across the land with enormous power. Nearly eight thousand cubic feet of water passed a particular rocky bluff every second, yet in times of flood, the volume could increase sixteenfold until the entire valley was inundated. More than two dozen floods of this magnitude had occurred in the last century, each time depositing sand and silt to depths of several feet upon the valley floor, enriching the soil in which the river forest grew.

The shoreline was varied by worn rocky outcrops, flat meadows gently sloping to the water's edge, and low sterile plains washed frequently by even minor floods. The river's meanders were mighty curves, although no more stable than those of its lesser tributary streams. It was now flowing faster than it had in most places upstream, for the retarding effects of friction from bed and banks were less in proportion to the huge volume of water passing down the valley. The greatest velocity ran a foot or two beneath the surface, away from the air which had its own slight braking effect.

Although there were no riffles, the surface was agitated by mysterious upwellings, streaming eddies, and miniature whirlpools caused by irregularities of bottom and current. Every rock and sunken log, every bend in the direction of the river and change in slope, created a churning mass in which water particles altered position constantly in relation to their neighbors and the stream bed.

Because of the massive turbulence, the river's temperature was uniform throughout. The water was not especially warm, for the brown turbidity absorbed much of the sun's heat in a thin layer at the surface, and light failed to penetrate very far. The warmth from the surface was being mixed constantly into the river, but because of the volume of water there was little change in temperature.

Extremely small silt particles, which caused the brown opacity, were carried along, tumbling through the water for nearly the entire length of the river, settling down to the bottom only when velocity diminished near the flaring mouth that emptied into a wide bay. Larger, heavier particles, with less surface compared to their volume and not so easily supported by the current, fell to the bottom long before the bay was reached. Because of complex variations in the rate of flow, these gradations were not rigidly defined: fine silt could be found in quiet backwaters of the upper river, while gravel might be carried down the central core of swift water all the way to the mouth and beyond. Boulders on the bottom decreased the speed of flow, usually accumulating deposits of fine sediment in the shelter of their downstream sides.

The river water by now had grown chemically complex, with many dissolved minerals from bottom sediments or from materials carried along in the current. Some substances were already dissolved by rain when they entered the river; others went into solution from plant fragments or stones that slid, bumped, and leaped along the bottom, being ground and pulverized as they went. In regions where bedrock was exposed, many deep potholes were scoured out by rocks now rounded to almost perfect spheres. The mechanical grinding liberated a wide variety of minerals into the water, and the pocked bottom created permanent and violent patterns of blending water and its contents. Such an unstable and violent environment made it quite impossible for plants and most animals to survive on the bottom. If they briefly took hold, they soon were crushed or wrenched from their grip.

Chemical nutrients that passed along in the water found only temporary lodging in some plant or animal, eventually being released and swept away again, either as wastes or when the organism died and decayed. The water and its contents were ever new and would not return again, so there was little chemical constancy in the river world. It could only reflect the sum of its contributors and had little character of its own.

The turbid river water was more dense and viscous than it had been upstream. Its load of suspended particles required much energy to drive them along, counteracting the pull of gravity; as a result the swift river moved a little more slowly than it would without such a load to carry. When a clear stream entered as a tributary, the dense, heavy brown water of the river immediately flowed beneath it until turbulent mixing made them indistinguishable.

The river environment changed daily, so all living things it supported had to contend with fluctuating supplies of food and dissolved gases. Although aquatic vegetation was generally scarce, a tapegrass grew commonly in the shallows. Only in coves and backwaters did the muddy shoreline support several kinds of plants that had flourished in the ox-bow ponds and smaller streams. Because the long ribbons of tapegrass provided shelter and feeding grounds close to shore, dace, suckers, black bass, and several other kinds of fishes swam slowly, heading into the current to maintain their position. In quiet coves indenting the banks, pondweeds streamed to the surface, covering the feeding grounds of small fresh-water clams and snails. Farther out, past the narrow zone of shoreline river plants but not as far as the great storming mass of river water, delicate, reddish worms and dark mussels burrowed. The worms were restricted mostly to flocculent silt and mud, while the mussels preferred sand and fine gravel.

Fresh-water mussels are clams, quite unrelated to the blue mussels of bays and rocky seacoasts. Mussels of the river were of several kinds, the most abundant being a rich dark blue or black, with white eroded portions near the hinge linking the two shells where the pearly layers beneath the dark outer covering were exposed. This tough covering, hornlike in composition, protected the animal's

shell from the battering effect of sand grains as well as a
slight acidity that increased when plant acids were flushed
from the surrounding land. It was produced along the
margin of a delicate fleshy mantle just inside the shell,
where it emerged in thin sheets from canals that were
twenty thousand times smaller than the limits of vision.

A clam dug its slow way through the sandy bottom with
a muscular white foot that could change from a blunt point
to a bulbous projection. Contraction of the foot, once it
was anchored in the sand, dragged the slim shelled creature
along from one location to the next. The animal dug into
the bottom at an angle. Emerging from the portion extending above the sand were two unequal openings: one,
fringed with fingerlike projections, inhaled water for food
and breathing, while the other, a simple tube, sent water
back into the stream. The current created within the clam
was due to millions of tiny beating hairs that covered the
surfaces of the mantle and the folded, curtainlike gills.
Oxygen from the river was taken directly into a simple
blood system, but tiny particles of food, mostly one-celled
plants, were collected on the surface of the gills in sheets of
mucus which flowed evenly toward the mouth. The mussel
was one of the most efficient trappers of microscopic life in
the entire river system.

Some of the mussels had swollen gills which were serving
as pouches for developing young. Quite unlike their parents, young mussels were enclosed in rounded transparent
shells equipped with tiny sharp teeth along the shell margin.
At present they were being shed by several parent mussels.
As they spewed forth dropping to the bottom or tumbling
along in the current, their tiny shells snapped wildly. Fishes
feeding nearby were attracted to some of the adult mussels
that extruded short lengths of waving mantle, simulating a
minnow even to the extent of having a black spot for an
"eye." When fishes neared the black clams, or swept close
to the bottom, fins dragging, larval mussels snapped on to

them and were carried away, or were lodged in a fish's gills, inhaled with water as it flowed through. Soon the tiny specks became encased in fish tissues where they would develop slowly for two or three weeks. When they had changed successfully into miniature mussels, they would emerge from the fins and gills of their host fish and, with luck, take up an independent residence on the bottom. The chance of reaching adulthood was slight, for flatworms and other small predators in the bottom attacked them almost at once. Even when they were older, large fishes, raccoons, and muskrats would break open their shells to get at the nutritious meat inside.

Out in the deeper, swifter portions of the river, life was nearly absent, but where light was available and the current lagged near the shore, yet provided enough oxygen and prevented silt from accumulating, animals and plants increased in number and variety. Where water flowed into the river from submerged springs, life was especially abundant; plants were well-nourished there and in turn provided food and shelter for animals.

Like their relatives upstream, the animals of a large river could not allow themselves to be swept away; they had to tolerate whatever conditions came their way, regardless of hour or season. Strong swimmers were able to migrate up and down the stream at will, but most river animals were neither large nor powerful enough to contend with the current. The smallest of sedentary creatures were fixed to the bottom, where friction slowed the water and they could feed and move about safely. Here were colonial rotifers, stalked one-celled protozoans, sponges, mats of moss animals, and brown hydras. Close by, snails crept along, and flatworms, leeches, and larval insects wriggled or scuttled across the bottom. Where the river bed was especially soft, burrowing mayfly nymphs joined the multitude of small worms in incalculable numbers.

Immature mayflies, delicate big-eyed creatures with fine

white gill filaments extending from their bodies, strained drifting particles of plant life from the water as their main source of food. With shovel-like legs and sloping, pointed heads, they easily penetrated the bottom, where they remained until, numbering in the millions, they came to the surface in spring and summer to emerge as short-lived flying insects.

A mayfly nymph, now ready to leave its river home, swam just beneath the surface, wriggling strenuously in the current that carried it downstream. During one convulsive spurt, its back split and the next stage in its life, a pre-adult, burst forth into the air, riding on the empty, buoyant skin while its wings dried. After several abortive attempts at flight, it finally took to the air and headed for shore, where it alighted on a bush and crept under a leaf to undergo its final molt.

After a few other mayfly nymphs had risen to the smooth surface of the river, the water suddenly began to froth and boil as the stimulus for emergence spread. For hundreds of yards there was nothing to be seen but the eruption and fluttering of pale, weak insects, sweeping cloudlike toward the forested shore. The sun glinted on a million wings and the air was filled with their rustling flight. Flycatchers, swallows, and other birds dived repeatedly into the swirling masses of insects; frogs sprang from their hiding places and nearly every kind of river fish rushed to the surface, where they noisily gulped one struggling nymph after the other. Yet the action of a multitude of predators seemed in no way to diminish the population of mayflies leaving the river. The water was littered with cast-off nymphal skins and adult carcasses which swirled and streamed down the central channel, or gathered in broad blankets in eddies along the shore.

For the next three or four evenings, the adults flew again on their mating flights, after which the males fluttered away and died. For a while the females flew vigorously upstream, rising and falling above the water, where they left clusters of eggs to be carried to the bottom; then they died too.

Although the river produced great quantities of food that supported life on the bottom—mayfly nymphs being the most commonly eaten creature of all—the great mass of the river itself was devoid of any permanent populations. Tiny planktonic animals and plants, common to ox-bow ponds and the downriver bay, were unable to contest the current and failed to establish populations that could breed safely in one locality.

The microscopic life of the river came mostly from lesser tributaries, flooded ponds, and permanent populations in the quiet bottom sediment, or it might be swept from tranquil weed-choked coves indenting the river bank. Drifting, planktonic life, mostly symmetrical green or yellow plant cells, was rare and on its way to an inevitable demise, separated from home populations with nearly all chance of survival gone. Even if the tiny lives successfully passed the rapids and falls downstream, they soon would feel the effects of salt from the sea, a condition they were not equipped to survive.

River plankton, such as it was, consisted of sparse populations of small crustaceans, rotifers, protozoans, bacteria, diatoms, green and blue-green algal cells. The bacteria, diatoms, and algae might reproduce on their way, at times creating a large and varied drifting population, but the same fate awaited them in tidal waters far below. A small copepod crustacean, carried away from its place of hatching, could cover as much as fifty miles in a day, totaling hundreds of miles during its short, two-week development. Only if it were carried into another cove or caught in a slow eddy would it have a chance of meeting more of its kind and reproducing. The ever-changing river was no place for a stationary plankton community.

It was the larger and more powerful swimmers that sur-

vived best—turtles, frogs, snakes, muskrats, mink, and
fishes. They were not obliged to fight the current constantly, for they usually sought shoreline waters, but their
senses were keen and told them when to head into the race
and swim vigorously to escape being carried out of control
downstream.

Swarms of small fishes, mostly black-nosed dace, hovered
in shallow water close to shore near the mouth of a small
tributary where they had been courting and mating for
over a month. A group of a half-dozen females hanging
close together in mid-water was soon disrupted by a male
which singled out one of them, darted into the assembly,
and pursued her toward the bottom. Immediately the chase
was joined by several other males, whose bright red fins at
this season contrasted with the dull colors of the females.
They gathered about, pressing headfirst against her body,
squirming and shimmering. Finally one male forced the
female to the bottom where they lay, his body encircling
hers, while she laid eggs and he released sperm over them.

This pursuit occurred widely along one section of
the bank, although it did not always terminate in mating.
No sooner did one male drive a female to the bottom, than
another would dive down and disrupt the process. The
dace which were not courting—over half the population
—swam singly or in schools, feeding on insects trapped in
the water film and finding bits of food on the muddy
bottom.

Suckers, light-colored fishes with fleshy lips that pointed
downward, were only slightly less common than the dace,
although there was no competition between the two. The
suckers were busy grubbing about the mud and sand for
larger insects and worms the dace could not eat, as well as
for decaying organic matter they found lodged between
rocks or wrapped around plant stems. These sluggish fishes
were important to the living economy of the river, for they

were caught frequently by larger predatory fishes, birds, and mammals that hunted in the shallows.

One of the smaller suckers strayed into deeper water where it swam in the swift current with difficulty. Before it could turn toward shore, a large, streamlined, blue-gray fish, rushing in powerfully from the open river, caught the sucker and swallowed it with one yawning gulp. Only a few weeks before, this salmon had ranged well off the seacoast, its body then a brilliant steel blue, seeking marine crustaceans as food. During the time it had ascended the river, its color and food habits had changed; even more importantly, it had become able to tolerate the fresh-water environment in which it now found itself. Unlike sea water, which had a chemical consistency somewhat similar to the bloodstream of the fish, river water was almost entirely saltless and passed all too easily into the tissues of the salmon. Suckers, dace, brook trout, and other permanent inhabitants of the river had well-developed means of eliminating excess water, but very few marine fishes were able to enter fresh water and survive: salmon and young eels were two of the rare animals along the coast which had the ability to do so.

The salmon, with tens of thousands of its own kind, was on its way upstream to spawn later in the summer. It would go as far as possible, to lakes and abrupt waterfalls too high to traverse. Then, its color lost, fins fleshy, and body distended, it would enter a new phase of existence preparatory to spawning. There would be much thrashing about and fighting among competitive males, using as weapons powerful lower jaws that had grown so long and curved that they could only touch at the tip.

The river contained salmon of all ages. Not only were large adults passing upstream, but some that had matured within the river itself were courting larger females just arrived from the ocean. There were young salmon hatched out only a few months before in the spring, others one or two years old and still not ready for the sea, and three-year-old salmon heading for the bay and open ocean. All were feeding voraciously on minnows and other small fishes. There were even a few huge and elderly salmon, at least a dozen years old, which had survived the annual spawning trip year after year, unlike their western relatives on the far side of the continent which died soon after their one time to spawn.

From the muddy sand around a submerged stand of tapegrass, a blunt gray head projected. Close to the tip of its nose were two pairs of elevated nostrils that sampled dissolved odors in the water, while a long, sharp-toothed mouth gulped water rhythmically. The water was exhaled beneath the sand from two fleshy openings on either side of the head, causing the sand to flutter slightly. As a sucker came overhead and began rooting in the bottom, a dark gray and golden eel burst forth from its resting place and with quick, sinuous curves, swam rapidly upstream where it once again settled beneath the surface. The eel, a female, had lived in the river for over a year, and would either remain there for five or six more years, or would migrate upstream to quieter waters, perhaps even wriggling over-land on a damp night to the protection offered by one of the ox-bow ponds. When enough years elapsed and she had grown large, black, and sexually mature, she would leave the ponds and streams which provided security and nourishment, and migrate downriver in the fall in the com-pany of thousands of other female eels. In the estuary and bay, they would be joined by males and together travel beyond the coast, across the broad shelf surrounding the continent, and head toward the deep sea far south into the subtropics, close to the region where the river had been born. There they would mate and die, but their young—curious, leaflike, and transparent creatures—would swim to surface waters and be carried along the great oceanic cur-

rent that passed northward along the continent. When a
suitable period of development had taken place, before the
current veered eastward to cross the northern part of the
ocean, the larval eels would head to shore and enter bays in
the springtime, now transformed into delicate, tiny elvers,
capable of making their slow way upstream to live and
mature. No inhabitant of the river, other than the salmon,
traveled so extensively to complete a life cycle, and even
the salmon did not exist in what almost seemed to be two
separate lives, neither resembling the other in appearance or
in habitat. The eel, a fish most unfishlike with its scaleless
skin and long, finned body, was both familiar and mysterious, hidden in the depths of the sea.

For an eel or a salmon to return to salt water from fresh
was no simpler an adjustment than entering a river system
from the ocean. As they entered sea water, less fluid entered
their tissues, but salts accumulated, and it was this danger
that denied access to the sea to most fresh-water animals. In
the salmon's case, its migrations were largely determined by
changing internal demands which it sought to satisfy. It
could not go downstream until it possessed a means of
getting rid of excess salt, a development that took place
slowly in its internal chemistry. The eel, on the other hand,
could go almost directly from one kind of water to the
other, changes soon taking place in salt-secreting cells in its
gills.

Not far from where the eel lay hidden in the sand, the
stately figure of a great bird flew across the sky, descending
toward the river shore. The huge, deeply curved wings of a
great blue heron suddenly began to sweep backwards while
its slender legs extended toward the surface. It landed
slowly and carefully, its legs sinking into a foot of water.
There the heron stood, motionless, neck stretched high and
rapierlike bill poised, its outline looming above the low
vegetation of the river bank. With stealthy tread, it stepped
into deeper water before freezing into immobility. The

heron stood without moving for half an hour, but when no moving prey came within reach, it resumed its slow and careful walk along the shore, keeping to water that reached almost to its blue-gray body. Now its neck was stretched forward, bill down, watching for a disturbance in the river. One of its broadly splayed toes touched the sand above the eel, causing the fish to burst upward and swim away from the bank, but the current slowed its momentum and forced it within a few inches of the surface. With a flashing thrust, the long neck of the heron shot down and the eel was seized in mid-body by an unerring bill. With a quick toss, the bird threw the eel into the air, deftly caught it headfirst, and swallowed it in two or three convulsive gulps, before resuming its slow and deliberate search along the bank, now without further success. The dace, suckers, and other river fish had been alarmed by the sudden lunge which had spread shock waves through the region, and all had fled to deeper water.

The heron leaned forward stretching out its neck, took a few quick steps, and with several especially powerful strokes launched itself into the air. Head pulled back against its shoulders, its neck forming a sinuous curve and long legs trailing, it swept across the river. On the flight across, it once hesitated, beating its great wings to hover directly above the surface, and extended its legs down to dabble in the water, exploring what seemed to be a dead fish being carried downstream just beneath the surface, but nothing came of the attempt and the huge bird continued on. As it passed over the river forest fringing the far shore, a flock of redwing blackbirds rose around the heron and harassed it, diving and milling about, causing the much larger bird to alter course and fly in dignified retreat over open water until it disappeared around a bend.

As afternoon waned, colors along the river bank grew warm and golden. Eddies in the water caught the low sun, sparkling and dancing in never-repeated patterns. Bright stems of emergent shoreline plants waved in the current, parting as a brilliant wood duck emerged, followed by its duller mate. They were feeding busily on insects trapped on the surface of the water, and upon others that crept upon the stems and leaves of plants. Whenever the supply of insects appeared to wane, they plucked at seeds and berries close to the water, and broke off bits of tapegrass and other submerged plants.

The dazzling colors of the male were about to be diminished by a summer molt, although most of the metallic blue and purple, bronze and green feathers would be kept, as would the bright reds and yellows of bill, feet, and eyes. In the spring the pair had nested high in a damaged white oak, fifty yards into the river forest. There the female had brooded a dozen eggs, deep inside a rotted part of the tree. Only a day after they had hatched, the ducklings had crawled up to the ledge of the hollow, where several were prodded by the female to fall fluttering to the soft leaf litter below. The rest responded to her low calls from the ground, and one by one they dropped, gently in their downy feathers, to land unharmed at the base of the tree. At once they had clustered around her to be led, single file, down the slope to the river. Now, several weeks later, the young wood ducks remained just inside the screen of shoreline plants, feeding in the same manner as their parents. They numbered only eight now, the others having fallen to predators such as the seven-foot pilot blacksnake that waited motionless at the base of a yellow willow only a few yards away.

Out in the water, a great splash shattered the quiet afternoon as a large salmon jumped clear, momentarily gleaming in the fading light. Its startling impact sent the young wood ducks running up the bank into a thicket, while the parents with harsh calls leaped into the air, flying rapidly and with great skill through the dense mid-levels of the forest, darting between tightly spaced branches without slackening speed. Soon they circled back, braked to a landing, and called the young ones from hiding.

When all was still, a pair of pointed nostrils appeared above the water several yards downstream, followed by a broad triangular head that gave insufficient evidence of the massive snapping turtle standing on tiptoes upon the muddy bottom almost two feet beneath. Its yellow-flecked eyes rolled back and forth while it pumped air into its lungs before slipping under the surface without a stir. Underwater the snapper was a graceful and powerful animal, stepping lightly over the bottom or swimming rapidly upstream, its long neck outstretched in anticipation of finding food. The turtle's sharp beak and muscled jaws made it a formidable predator, but the greatest quantity of food it consumed proved it to be a vegetarian and a scavenger, devouring weakened fish and aquatic carrion. In its fifteen years of life, it had succeeded in eating very few young waterfowl, and only when they had paddled unsuspecting within its striking range. The snapping turtle was almost exclusively an underwater feeder; in fact, it was so aquatic it seldom basked on the river's edge, although the females had emerged fully to lay eggs high on the bank in late spring.

The snapper had not been watching the wood ducks, but was intent on a diseased and dying leopard frog thrusting itself feebly along the shore. The turtle crept toward it, head cocked back on its muscular neck and hind quarters elevated. Within range, it stopped for a moment. Then it was still until, with a burst of enormous power, it leaped forward; its head shot out and the frog was clamped between mighty jaws. Part of the victim was swallowed at once, while the hind quarters were shredded by the clawed front feet of the turtle and held until another bite could be taken.

The weakened frog had been noticed by other eyes as well: lying on a log caught in the shallows, a banded water

snake had coiled in preparation. But the explosive burst of
the snapper caused the rich red and brown snake to slip
quickly from the far side of the log and swim smoothly
upstream. Removed from the scene, it wrapped around a
clump of pickerelweed to keep from being swept back
with the current. Its brown head poked above the surface
where its lidless eyes watched for more activity.

As shadows crept out from the western shore, reaching
toward the gold-tipped trees across the river, the surrounding forest was filled with new sounds—the songs of tree
crickets, frogs in the wet swampland margins, and the
booming, cackling, prolonged hoots of barred owls. An
owl concert was beginning, with more of the large birds
flying along the valley with slow wingbeats, silhouetted
against the fading sky. Dozens came on quiet wings, settling into the trees, until they occupied a mile of the river
forest. Their reverberating calls, each distinguishable indi-
vidually by pitch and syllable, created a cacophony. These
birds, which hunted mice along the shore and caught fish
and crayfish in shallow water without difficulty, often
gathered in the early evening hours to call back and forth,
perched amid the densest growth of trees and vines. The
loud, emphatic hooting of one owl would be answered
immediately by another, then picked up by others, until the
river forest was alive with deep, rhythmic, accented calls
which began slowly, rose in pitch and intensity, then diminished to a harsh cough. The wild, loud sounds were rich
in variety, at times interspersed by a rapid series of cackling
and whooping cries. All other animal voices were lessened
by the din, and mammals of the area kept under cover or
left the region where the owls had concentrated.

For many of the smaller mammals an entirely different
sort of signal caused them to restrict their nocturnal travels.
It was an acrid, penetrating odor that followed the waterline, the secretion of a mink walking slowly along the bank,
long body arched high above its short legs. The mink was a

male seeking food for its six lightly furred young that
nestled in their den under the roots of a willow overhanging
the river shore. The den, an old muskrat hole, was padded
with leaves, feathers, and other soft material, and was littered with the remains of numerous kills. In one corner was
a stockpile of several mice, a muskrat, a young wood duck,
and an ovenbird, all freshly killed. Both parent mink
hunted in the dense woodlands of the valley shore, keeping
close to the river margin where they often took to water to
capture crayfish, mussels, and fish. Although the eyes of the
young mink were not yet open, they already displayed
their carnivorous nature by chewing on fish and mice their
parents brought back.

The mink paused, then broke into a rapid, bounding run
and plunged into water only inches deep. It emerged instantly with a young bullfrog in its jaws; a quick bite
behind the frog's head, and the frantic kicking ceased. The
male turned toward the den, which was close, and slipped
inside where the young mink were already squealing with
hunger. They immediately fell to mouthing and chewing
the moist, tough flesh of the bullfrog—not that it served as
food yet: they would have to await the female's return
from her woodland hunt before they could be nourished
by her milk. The cache in the corner, which the young
mink had been unable to find, was a stockpile for a day or
two hence, when their eyes would open and nursing would
cease. The demands placed upon the parents until the
young were capable of hunting for themselves would be
heavy indeed, and there would be no cessation to the
brood's growling, hissing, and ferocious play until that day
arrived. Later, after they had learned to hunt with their
parents during nocturnal journeys, the family would separate to adjacent or overlapping territories up and down the
river shore, leading solitary lives until they sought mates
the following spring.

On a far smaller scale and quite overlooked by the male

mink, a tiny water shrew was busy scrabbling about on the
sand and mud bottom close to shore. It glistened with a
silvery sheen in the intermittent moonlight, for its hair held
a coat of air so effectively that the skin beneath was never
wet. This made it buoyant, so it held onto pebbles and
plant stems while searching for insect larvae, fish eggs, and
small fishes that had grown quiet for the night. The water
shrew had one of the most intense rates of living of any
animal in the entire river system; to sustain a rapid loss of
heat from its tiny body and a heart that beat nearly eight
hundred times a minute, it was obliged to eat constantly.
Even so, it would not live much more than a year, in spite
of avoiding its many enemies which included herons, mink,
and predatory fishes, for this very intensity of life meant
that it aged more quickly than any other small mammal. It
would die of old age in the early part of its second year,
having in the meantime raised fifty or sixty offspring, few
of which would survive even briefly to pursue their frantic
way of life.

Through the night, the river flowed on, mile after mile,
widening its course and conveying more water. The high
hills were left far behind; now there were only gentle
elevations and they too were flattening out. The river bed
was still one of rock, but it was frequently covered by
muddy sand and gravel which in places rose to the surface
to form sand bars and long, elliptical islands anchored by
shrubs and grasses. The banks were nearly a mile apart, low
and indistinct in the diffused light of a full moon increasingly obscured by low-flying clouds.

As the night wore on, the light grew dimmer; a flickering across the northwestern horizon foretold the approach
of a thunderstorm coming down the shallow valley. By the
time the massive, towering cloud formation was twenty
miles away, a deep rumbling filled the air as whole areas of
the cloud bank were brilliantly lit again and again. The
atmosphere over the river grew still, almost oppressive, as

the storm continued to build inland, its cold dense air slipping down the gentle slopes, tending to follow the depression created by the river.

By early morning hours in what normally would be twilight, the thunderstorm arrived, preceded first by strong, chilly winds, then accompanied by sleeting, blinding gusts of rain mixed with hail, limiting visibility to less than fifty feet. Waves and ripples were flattened by the enormous deluge as rain pocked the surface and millions of leaping, spurting drops rose and fell across the expanse of dark water. Where rain struck sand along the shore, countless pits sprang into being, looking like the energetic activity of subsurface worms digging everywhere, entrances to their tunnels collapsing one after another.

Low ragged clouds raced down the wide river, the blinding light of repeated lightning now illuminating the whole valley with bright, blue-white flashes. The explosive booms of thunder gave way to sharp, jarring cracks, their concussion so great the ground and trees shook with renewed fury. Gusts of wind hit the shore at seventy-five miles an hour, tearing up bushes and splintering trees. In the cold, turbulent air above, huge vertical drafts whipped leaves, dust, and water droplets to thirty thousand feet in five minutes, then drove them down again at equal speeds. The drops of water were primarily responsible for electrifying the air, partly by friction, but especially when each drop was broken or atomized in the fury of the storm and its fragments took with them opposing electrical charges. More rain and hail falling into such electrical fields captured these charged particles and assumed their polarized nature. Because of the violent updrafts, many of the negative particles accumulating in the lower regions of the storm were driven up again, becoming neutral or even positively charged if they were carried high enough into the towering clouds. The whole core of the storm contained prodigious quantities of harnessed solar energy, released by condensation of water and now being converted into violent winds and uncontrolled electrical power greater than any other localized natural force on earth.

Just before the discharge of a bolt of lightning, the energy gradient directed downward at the earth reached nearly sixty thousand volts an inch, suddenly released in thirty or forty distinct strokes of lightning, each separated by only hundredths of a second. The first, most powerful stroke branched out, its luminous dartlike advance hesitating on its way to earth, seeking the path of least resistance. Subsequent strokes rapidly pursued the most effective path pioneered by the first, each leader immediately followed by an intensely bright return stroke upward from the surface. The violent release of pent-up energy was done in a second or two and the sky was momentarily black until the next discharge burst forth.

Waves began to pound against the river banks, gusts of dirty spray flying inland. Tiny minnows were thrown upon the gravel shore and washed out again, almost invisible in the turbid water among quantities of debris and remains of plant stems. They tried to zigzag along the shoreline, now heading into the bank as a wave receded, now heading out as a wave rolled in, always trying to maintain their position between shore and deeper swift water but often failing. A large fish ripped through the battered school of minnows so rapidly they did not appear to notice it, although several were snapped up.

The wind began to diminish and the heavy rain lessened; light spots grew in the torn clouds that fled after the departing, grumbling storm. The surface of the river grew calm, although muddied by soil washed in by the downpour. New sediment streaked the central channel, running between ribbons of darker and more reflective water. The violence had passed, leaving little change, for all the water spilled into the river by the storm had been carried away at once with only little effect upon its level.

Patches of clear sky opened and high pink billowing clouds, lighted by the rising sun, appeared to the northwest while the opposite horizon remained darkened by the impenetrable bank of low storm clouds, haloed along their outer margins. The air was cool and suddenly dry. Flies and gnats, buffeted by the wind and rain, buzzed out from the dark havens of shoreline undergrowth where they had taken shelter.

So much water passed through the broad channel, and so rapidly, that it had little time to affect the low land of the rocky piedmont over which it ran, except during times of spring floods. There were now two distinct worlds: that of the great river plunging toward the sea, and the land which stretched to the horizon beyond either shore. There were fewer transitional animals present; they were either thoroughly aquatic or restricted to life on the land, both kinds avoiding the banks that were almost devoid of vegetation where the velocity of the water was too swift for most organisms to find anchorage and security. The close association between river and land that had existed for hundreds of miles was nearly gone, with the river rushing straight down the gentle slope, hurrying toward its return and dissolution in the broad ocean.

VIII

THE TOUCH OF
THE SEA

*Nature . . . is but a succession of
changes so gentle and easy that we
can scarcely mark their progress.*
CHARLES DICKENS

T H E river drew near the edge of the highland piedmont, the ancient continental shoreline that a million years
ago had ceased to be the boundary between land and sea.
The great ice caps of the north had advanced four times
and four times retreated, leaving rivers and streams to carry
away debris from their rough passage over the land. Deltas
formed around the rivers' mouths, so that a thick and wide
mass of sediment stretched out from the continent, two or
three hundred miles east of the original coast and extending
its entire length. Depending upon the amount of global
water locked in the distant polar regions, the shoreline had
fluctuated back and forth across this gently sloping sedimentary plain. During times of extreme glaciation, when
water remained as ice instead of returning to the ocean, the
shoreline, considerably below its present level, had extended far out on the coastal plain and rivers fell directly
into the sea. In warm periods midway between glacial ages,
the sea level had risen a hundred feet above what it was

now, invading rivers and creating estuaries, and waves
again had broken close to the original shoreline. The present seacoast was about halfway between the ancient river
mouth and where the accumulated sediment dropped off
steeply toward the ocean floor far below, but was creeping
slowly inland as the ice caps gradually melted a little more
each year.

The estuary had not always been divided into a tidal
river and a wide, bell-shaped bay. All estuaries are ephem-
eral geological events, rare in the history of the earth. As
the last glacial age came to an end, when the ice caps still
lowered the sea level, the estuary had been a huge, narrow
river that took the brunt of melt-water directly to the
ocean. When the glaciers retreated, new rivers developed
to the north along the coast, relieving this one and allowing
it to grow quieter with a decreased flow, although a few
others joined it from time to time as they eroded different
channels in the mountains and the piedmont. Sediments,
from rocks to the finest silt, were deposited in ever-increas-
ing quantity. By now the rate of melting had slowed, but
the sea level was still rising two feet a century and would
continue as the ice caps diminished until sometime in the
future when the first twinges of a new glacial epoch would
be felt. As the sea level rose, the lower river valley began to
be flooded, until six thousand years ago it had widened into
a shallow bay, superimposed upon the old river which had
been invaded. So the breadth of the bay would go on
increasing, possibly joining others along the coast, at present separated only by low peninsulas. Eventually, with the
world in the grip of a new ice age, all estuaries would
disappear again and much of their specialized life with
them. Clues still existed in the present estuary revealing the
course of the old drowned river, for most water flowing
down followed the original channel that hugged the western shore of the bay. The entire eastern side was a vast

shallow basin well flushed by tides, but little influenced by the river.

The river had run its present course over hard rock of the higher piedmont for several thousand years, cutting only a shallow bed in the valley floor. But where it left the old continent and began flowing across sediments of the coastal plain, the rushing water had a profound effect, scouring out a deep channel without difficulty. At the precise point of departure from its old, hard bed to the lower, softer one, the river descended abruptly in rapids and a series of falls that churned the water to violence. The extreme turbulence did not last long; soon the river flowed placidly across the flat and featureless coastal plain. Yet it was here, in this hundred-mile length of estuary until it reached the sea, that the most complex events of the river's entire existence took place. Nowhere else in the world of water was there a region of such importance to life in the sea, nor one with such a multiplicity of environmental factors.

Past the falls, banks of the river estuary rose forty feet above the water, layered in gray and red clay, and a light sand, which were sediments deposited during the formation of the coastal plain. Great chunks of bank soil had collapsed into the water, creating conical talus slopes. The narrow beaches were littered with stones and tree trunks; a few trees had fallen from the heavily forested brim and continued to grow at the water's edge. One tree hung upside down, still rooted at the crest of the bank, growing and producing foliage twenty feet below. Vines followed crevices and rills that intersected the layered soil. Dense green thickets terminated abruptly at the edge of the small cliff, except where turf mats spilled down into concavities left by the collapse of loose earth underneath. In some places, whole sections of the mat had broken off and grew, not too successfully, at lower levels.

These hundred miles of estuarine waters, rising and falling, clearly responded to the tug of the moon during its
monthly cycles. From the surface it still appeared to be a
broad river, for the complexity of its movements and the
events that took place were hidden within its enormous
mass. The estuary fed the ocean and its life; it was a
nursery for multitudes of marine creatures and plants, and a
refuge for others; it was bordered by some of the richest
lands on earth; and it was a multidimensional changing
environment that supported some of the most specialized of
inhabitants and densest populations.

By now the estuary was not only the sum of all the
river's tributaries, but was augmented by tidal masses to
such an extent that a half-million cubic feet of water would
pass by every second, either entering or leaving except
those few minutes when a diminishing incoming tide would
balance the river flow, bringing the whole estuary in one
region to a virtual standstill.

The water level at the foot of the banks was not constant. Twice daily the flow of the river was retarded by the
pressure of tides rushing from the open sea all the way to
the falls, and twice daily the yellowish downstream flow
was accelerated by departing tides. Because of the compression of a tidal surge into the funnel of both bay and narrowing river, the eight-foot rise along the shores of the
upper estuary was twice as high as that of the seashore,
creating conditions along the banks that made life for attaching organisms physically difficult. It was such an isolated and restricted environment in the nature of its characteristics, fraught with profound change, that no sessile
intertidal creatures had ever succeeded in becoming populous at the head of the estuary. The plants and animals that
were present either were from riffle areas upriver, or a few
hardy migrants from the bay.

Particles of water that had descended the whole length

of the river without interruption, save an occasional eddy
or backwater, now were caught in this tidal oscillation that
impeded their progress. Instead of flowing four or five
miles an hour, it might take river water three months to
travel the hundred miles from falls to the ocean beyond the
bay. The tidal surge back and forth established a degree of
permanence to water in the estuary compared to that of the
river above the falls. But this same tidal flood and ebb, with
velocities as great as four feet a second, scoured the bottom
of the upper estuary and made attachment by plant or
animal very nearly impossible.

Another major development, not present in the upland
river, was the gradually increasing salt content as entering
sea water was dissolved in fresh. Near the fall line at the
head of the estuary, the faint tinge of salt was so slight that
it had little effect upon fresh-water animals living below
the falls, and allowed only a rare invader from the bay to
survive for any length of time: an occasional blue crab or
hog-choker flounder would appear riding on a flood tide,
but usually would return downstream to a lesser dilution
before long. A rare brown-stained barnacle was attached
here and there to an old log or rock along the shore. These
few barnacles had been carried up the estuary during their
planktonic larval stages, encased safely but temporarily in a
small bubble of sea water, surviving long enough to attach
to a permanent surface. Once fixed in place, they remained,
with shells and valves able to close off the watery environment whenever it became unfavorable.

Although estuaries are only temporary features in the
geology of shorelines, they have played an essential role in
the development of life, quite out of keeping with their
short span of existence. During the earth's early history,
estuaries had been major invasion routes to fresh water and
eventually to the land. Estuarine systems in the area had
developed at least one hundred sixty-five million years

ago, leaving faint traces in present-day rocks and sediments, but all record of preexisting estuaries had vanished completely.

Few animals or plants had ever evolved sufficiently to leave the sea directly for a terrestrial existence. Nearly every form of life now on land had originally emerged in an ancestral state from fresh-water ponds, swamps, and river headwaters as successive generations gradually had made the transition from open water to moist soils. Certainly this had been the route of all mammals, birds, and reptiles from their fresh-water amphibian and fishlike progenitors which had developed the ability to breathe air before flopping from one shallow, drying pool to another. It had also been the route for insects and their terrestrial millipede ancestors, evolved from some distant, aquatic predecessor. Flowering plants and ferns now covered the land, but mosses were still held to wet soils and shady regions where the drying effects of the sun were negligible. The developing spore of any moss resembled algal filaments in ponds and streams, some of which in turn branched profusely and even developed simple leaflike structures not unlike those of the mosses. The separation between aquatic life and that of the land was often indistinct.

Some of these migrations had occurred in the distant past; others were taking place now. In the estuary below the falls, wherever the water slowed near the mouth of a tributary creek, the bottom was heavily populated with small, humpbacked glass shrimps, transparent but for their eyes and brown digestive tracts. Their delicate, spindly legs probed for nourishment in the sediment, while long, hair-like antennae flicked about, sampling odors, currents, and the proximity of their own kind. They penetrated far up creeks and lived without difficulty in small fresh-water ponds that had been created by meanders or a slumping of bank soil sufficient to create a dam. That the glass shrimp was a form of life well on its way into fresh water was made clear by the presence of other distinct populations of its kind, only slightly different in form and tolerance, concentrated farther down in mid-estuary, in the bay, and even outside along the seashore. The long, slow invasion was nearing a successful conclusion, made possible only by continuous slight variations in body chemistry during the reproduction of these small crustaceans. In every generation there had been some individuals a little more successful than others in penetrating upstream, and they in turn handed down their traits to their offspring. They were prolific creatures and many females carried beneath their arched abdomens hundreds of eggs, some of which had developed two tiny black eyespots. Other eggs had already transformed into minute shrimp, nearly ready to depart into the plankton or the sediment where only a few would survive capture by ubiquitous predators.

There were times when this part of the upper estuary teemed with life. Only two months before, when the water had been warmed by the spring sun, millions of heavy-bodied shad had left deep water off the mouth of the bay and migrated into the fresh-water region below the falls. There they scarcely fed at all, but were so intent upon spawning that the water appeared to boil with their activity as silvery bodies flashed in the sunlight. The multitudes of young shad that had survived this far in their development, by now were feeding voraciously on aquatic insects and small crustaceans. They would remain in tidal creeks and the estuary itself until fall, when they would migrate to the sea to grow slowly for three or four years before again returning to a river environment to spawn.

This first part of the estuary had lesser populations and fewer kinds of life forms permanently established than the river above. Populations would grow even sparser for the next ten miles until the frontiers of the prolific bay and sea beyond were approached. The few silt-free surfaces exposed beneath these estuarine headwaters had almost mo-

notonously reduced communities of very small plants and
animals. Attached or surface-dwelling diatoms were plentiful, but did not live in great variety. Among them were
delicate and beautiful clusters of stalked, single-celled protozoans, each colony consisting of many hundreds of bell-
shaped cells branching off a common stem. They filtered
the water for bacteria, but when a larger object came
bumping along in the current, an entire colony would
retract strenuously into a tight mass. Seconds later the
contractile fibers within the stalks would relax and the
colony unfurl, each cell finally opening its bell to the
water, tiny fringing hairs beating to create a vortex that
carried particles of food into a curved gullet. While this
was a common type of animal on washed, nonsilted surfaces, the muddy bottom was penetrated by large numbers
of the same burrowing worms so common in the river
above the falls.

The upper estuary differed most from the river in its life
of the open surface where tidal water masses flowed back
and forth, and plankton had a chance to survive. The
plankton that were present were not so much ordinary
river types, but true estuarine forms existing throughout
the length of the tidal river and well out into the bay. One
reddish, energetic copepod crustacean was so abundant, it
gave the water a distinctive rusty color, clearly distinguishable from the yellow-brown sediments being carried along.

The course of the estuary ran directly to the bay, with
only a few major turns in direction. Meanders were
straightened out, for periodic incoming tides strongly reversed the effects of river flow. Sand was a common ingre-
dient along the shoreline, usually interspersed with silt and
gravel, and at times building up into exposed bars. In some
spots, there were even sizable glacial boulders that had been
rolled downstream by spring floods before coming to rest.
Heavy clay outcroppings appeared, interrupted by stones
and rubble. The banks could be steep and abrupt, yet else-

where sloped gradually inland with high-tide lines a hundred yards removed from the lowest water level. Rushes —spiked grasses with little brown flowers toward the end of each erect shaft—were the dominant plants along the shoreline, where they were inundated by the highest tides.

As the estuary turned and progressed almost due south, widening toward the bay, it approached a region where the range of changing conditions was greater than at any other time or place in the river's history. Near the head of the bay, every feature of the estuary was in continual flux. River flow, now growing unconfined, lost velocity and dispersed, releasing much of its sediment over a wide area. Incoming tides, on the other hand, amplified by the effects of severe compression as they entered the narrowing estuary, lifted river sediment into suspension again, creating a widespread turbidity. Here too, the fickle temperatures of the river met the more stable ones of the sea, producing wide variances depending upon the state of the tide and the season.

The quantity of water passing through the bay head and lower estuary was subject to extreme fluctuations daily, monthly, and seasonally. In spring, the river flow might exceed that of late summer by two hundred times; every month there were exceptionally high and low tides when the sun and moon pulled in concert; and there were the regular twice-daily tides—three completely separate cycles superimposed upon one another.

While these events could affect plants and animals attempting to survive in an essentially unstable region, there was still another factor that posed the most serious limitations of all: the constantly changing salt content of the water. In one thirty-square-mile area at the head of the bay, the water could be either as salty as the open sea, or nearly fresh. What was essentially sea water could penetrate twenty miles into the river estuary or recede far down the bay, depending upon the forces of the river combating

those of the sea. Such extremes, reflections of varying contributions from the river during severe spring floods or
prolonged late summer droughts, naturally brought about
major changes in the life present.

The separation between fresh and salt water was not
abrupt, but was complicated by differences in density between the lighter fresh water of the river estuary and the
heavy salt water from the sea, thrusting in and out twice
every day. On an incoming tide, sea water penetrated farthest upstream along the bottom in a wedgelike fashion,
while the river flow passed overhead. A member of the
fresh-water plankton from upstream could pass far down
the bay at the surface, while a marine invader, crawling
along the bottom, might find perfectly satisfactory conditions many miles upstream. The tidal wedge allowed salts
to build up in river bottom sediments, so marine microbes
and worms favoring life in the mud lived safely far up the
estuary. Conditions encouraged the growth of microbes,
for they lived in the estuary more abundantly than in either
sea or fresh-water river.

The bay was so vast—five hundred square miles—that
forces developed by the rotation of the earth affected it.
The southward-flowing river, variable and turbulent, was
deflected along the western wall of the bay, while incoming tides swept into the eastern side, setting in motion a
huge counterclockwise current from the wide shallow flats
of the east toward the deep channel along the west side of
the bay. This meant that the eastern part of the bay was
usually a little more saline than the western channel region,
so there came to be differences in the established populations of either side.

The life of this inconstant bay, plant or animal, was of
two basic kinds: those which survived only so long as
conditions met their needs, and life that actually flourished
in a widely fluctuating environment. The former had
come and gone innumerable times during the thousands of

years the estuary had existed, always being repopulated by familial stocks in the sea, river, or upper estuary. Those few that thrived on change, however, were the safest of all and survived the longest. If they were tolerant of widely different dilutions of salt water, they often were able to withstand seasonal temperatures and other conditions. They were, in fact, adapted to instability of the environment, for change was a permanent feature of the estuary. The success of an estuarine organism was seldom limited by a sudden alteration in a single environmental condition; usually it took an interaction of several factors to bring about the demise of a plant or animal. But when a creature vanished from the estuary, inevitably it would reappear later—the next season or even years after—its ability to repopulate the region depending upon season, its means of migration, and its potential for providing more of its own kind.

Communities of organisms were rarely clearly defined. They were composed of associations that waxed and waned and interfused with one another, each form of life having selected specific zones according to its needs of salt content, warmth, light, oxygen, current, type of bottom, and especially its ability to adjust to fluctuating and unstable conditions.

As a zone of transition between two far more constant environments, the estuary demanded so much of its inhabitants that often their physical structure and behavior were altered, as well as their ability to adjust chemically. Some found a kind of race security in enormously productive breeding, despite high mortality. Others became adjusted to riding the tides in such a way that they were confined to the bay. Some developed structural specialties, resistant stages in the form of spores, cysts, or protected eggs, or a means of migrating to escape unfavorable conditions. There were those which retreated into burrows or shells that could be sealed tightly; some were able to withdraw their most sensitive body parts; a few were capable of

secreting protective substances around their bodies. For
many, the tolerances or physical adaptations had come
about slowly through inheritance, until now these animals
were restricted to estuarine conditions and no longer could
return to either the sea or purely fresh water. Organisms
attached to the bottom had to take whatever came their
way, or die.

Where river estuary joined the bay, salinity differences
were greatest. It was here that salt or the lack of it was most
likely to be a limiting factor and life was not so varied.
Most of the organisms present had originated in the sea; less
than a dozen of the hundred different kinds of animals had
emigrated from fresh water. These, mostly insects, owed
their existence in the estuary to a thick, impervious body
covering which diminished the passage of salts into their
tissues, a development that also allowed turtles and muskrats to penetrate the bay from their river origin without
harm. One insect, a midge larva, had developed such
efficient salt-secreting filaments on its body that it continued to exist all through the bay, even on rocks and logs
at the cape itself. A few fresh-water fishes—white crappies,
bluegills, and bridle shiners—lived close to the shore in the
upper bay where seepage of ground water caused further
dilutions.

For marine animals in the headwaters of the bay and the
river estuary, the problem was one of avoiding an excessive
inflow of water fresher than their own body fluids. Because
large size usually meant an increasing inadequacy for removing fresh water that seeped into their tissues, animals of
the area having a marine origin were mostly small. The
same sandworm that lived more abundantly in the lower
bay, was here only one or two inches long compared with
the eight to ten inches it reached in more saline waters.
Small fishes, such as the mummichog, had more extensive
gill surfaces in relation to their bodies than the much larger
fishes of the lower bay, because gills were used not only for

breathing but as excretory organs. The little mummichog and its killifish relatives, the most abundant of fishes in this changeable region, swarmed through the turbid, warm waters close to shore, often rippling the surface as they contested the current. They fed upon whatever small worms and crustaceans they could find, but were far from being exclusively predatory: their stomachs were crammed with algae and plant remains that the tides had swept from the shoreline.

Not all creatures of the upper bay and estuary were small. Several enormous sturgeons, the largest at least ten feet long and weighing over five hundred pounds, swam slowly over the murky bottom, their sensitive lips and drooping, fleshy barbels in contact with the muddy sand ready to detect food. One male suddenly dove downward, pushed its heavily armored blunt snout into the soft bottom and began rooting about, shoveling up clouds of mud and debris. It kept dislodging small rocks to which it paid no attention, then struck a bed of soft-shell clams. Its mouth, far underneath the head, opened funnel-like with projecting lips and sucked in one clam after another. The other sturgeons, detecting the presence of food, immediately began digging into the bottom, until the whole area was a swirling, opaque cloud of brown silt.

These great primitive fishes, still closely resembling some of the most ancient fish ancestors of five hundred million years ago, had heavy, protruding bony scales set like tiles into their skin. They had even degenerated from their ancestral types, for now their internal skeletons, no longer able to produce bone, were completely cartilaginous. But absence of a rigid skeleton and the presence of heavy, bony plates set into their tough skin did not make the sturgeons sluggish fishes. During their feeding frenzy, several jumped clear of the water, returning with a resounding crash that crossed the water and echoed from the low cliffs.

The sturgeons were on their way back to the sea after

having spawned far up the estuary a month earlier, where each female, fifteen years old or more, had laid over two million eggs.

The tide was ebbing at present and water that had slowly worked its way down the estuary now raced into the bay, joining great masses of sea water draining from the eastern flats. The entire shallow eastern side of the bay was floored with a hard, rough bottom, supporting almost two hundred square miles of oyster beds. If there was one characteristic bay community, it was created by millions of oysters growing on the bottom, fixed to any available surface. Their very presence established a vast, interdependent association of plants and animals which otherwise would have had great difficulty in existing in the bay.

There were many active, migratory creatures in the bay, but the oyster was not one of these except during its larval days. Despite a sedentary nature, it and its ancestors had been the most successful and typical animals of estuaries and shallow seas for over eighty million years. Ancient oysters had been more symmetrical and clamlike than those of the present; their shell remains were revealed where the banks of the bay and estuary had been eroded by strong currents. Some of the more recent layers, dating back thirty-five thousand years or so, contained fossils indistinguishable from those oysters presently populating the bottom of the shallow flats. Similar layers were still being deposited, for beneath the wide beds of living oysters was a thick layer of shells that had settled into the bottom and been covered during recent centuries.

The oysters of the bay varied greatly in shape and, except when they were young, displayed little of the symmetry of their ancestors. If an oyster found a clear, uncongested spot on which to grow—this was a rarity—it developed a fairly rounded, fan-shaped shell. But most of them crowded together in contorted, cemented masses, so their shells were elongated and often twisted. In places where they had to grow upright from older ones now buried in sediment, the shells might be six or seven inches long, only two inches wide, and in this way reach above the suffocating silt of the bottom.

In its basic structure, an oyster resembled the fresh-water mussel of the river far upstream, but was so specialized for its life in the changeable estuary that it appeared to be quite different. It fed on diatoms, bacteria, and other living cells in the same fashion, creating strong, steady currents across its gills where millions of tiny hairs beat in synchronized waves. Each gill, supported internally by stiffened rods and as much a respiratory organ as a feeding device, was a highly organized trellis of holes and channels, down which trapped food flowed toward the mouth. In the vicinity of the mouth, food was sorted by delicate leaflike palps, bulky material being discarded. Only a small part of the possible food was swallowed, yet even this amounted to a great deal, since each large oyster filtered thirty quarts an hour when there were no sudden changes in the region to bother or alarm it. Sediment particles and excess food were trapped in mucus on the gills and palps and discarded, flowing out between the shells where they dropped to the bottom, enriching the environment for a multitude of small and microscopic lives.

Although an oyster was fixed to the substrate, it had not lost the ability to be internally active. Muscles regularly contracted in the gills, flushing blood back into the circulatory spaces of the body and allowing more blood to flow out where exchange of oxygen and carbon dioxide could take place. Its upper shell periodically opened and closed, controlling the passage of water. When a crab walked overhead, or a fish began browsing on some of the plants and animals attached to the oyster, it slammed its shells together in less than a thousandth of a second. That it could be bothered, indicated a certain sensitivity. There were special cells along the edge of the enveloping mantle that could determine shadows, strong light, temperature, and chemical changes, as well as the pressure of something else touching it. The single muscle that operated the hinged shells actually was composed of two parts, the smaller portion used for quick response. The bulk of the muscle was incapable of instantaneous reaction, but once closed, it held the shells together with a prolonged and powerful contraction. The muscle was so important to the welfare of the oyster that it constituted at least a third of its total weight.

The success of the oyster in the estuary was not only due to its wide tolerance of estuarine conditions, ranging from near sea water to water only one-sixth as salty, but to its great fecundity, and the way in which its larval stages took advantage of circulation patterns in the bay. Now that the water had warmed to over seventy degrees, nearly all of the adult oysters on both sides of the bay were spawning.

Initially it was the males that began the process. With shells gaping slightly, they periodically discharged clouds of sperm with exhaled water into the current which swept across the bottom. The success of the whole spawning venture was dependent upon the closeness of one oyster to another, for not only could sperm and eggs be lost if they were too widely spaced, but the release of sperm stimulated both other males and females to emit their sexual products.

In the females, eggs had been accumulating between the gills and the mantle, the mantle sealing them in by pressing its edges together. As a small aperture was formed between the edges of the mantle, a mass of eggs appeared in the opening. Immediately the large muscle of the oyster contracted violently, shooting out streams of eggs into the water. This was repeated several times until the supply was temporarily exhausted, after which the female was obliged to wait several days before being able to spawn again. Throughout the spawning season each female would produce many millions of eggs; during particularly intense periods, long, wispy streams of oyster spawn would drift along in the water.

Females now releasing eggs had not always been females. Earlier in their lives they were males, sperm production inhibiting the manufacture of eggs. Then for a period, they were bisexual, producing both sperm and eggs before finally settling down to egg formation. But this was not final, and often the females would turn back into males, their roles reversing at times as frequently as once a year.

With both eggs and sperm in the water, the eggs produced a chemical substance which caused the sperm to clump together, making fertilization of any one egg more likely. Each sperm, a tiny, expendable cell set into frenzied motion with a long whiplike tail powered by a limited amount of stored energy, thrashed violently ahead. In the proximity of an egg, a sperm cell usually thrust out a short filament from its head, the latter being the nucleus which contained coded hereditary material. The filament aided in chemical penetration of the egg cell, whose tightly surrounding membrane almost immediately inflated outward, denying further entrance by other sperm cells. As a successful sperm entered an egg, its dance done, the tail became still and detached. Once the nucleus of a sperm joined that of the egg, a new life immediately came into existence, although its chance of reaching adulthood was minimal. Within minutes, the spherical fertilized egg began to show signs of dividing into two hemispheres, after which they in turn divided. The stored nourishment in the egg served long enough to produce a tiny, oval larva, its two delicate shells joined by a straight hinge.

During the next two weeks, the millions of larval oysters added to the plankton of the bay would go through a series of distinct stages, slowly changing in form and size, relying upon structures they would later lose as adults. From between the two shells of each protruded a swimming membrane, fringed with long cilia beating in synchrony to propel the tiny creature through the water. It was also essential in gathering food, for along the base of the membrane were smaller cilia that swept the most minute of planktonic food into the mouth of the larval mollusc. When necessary, the membrane, or velum, could be retracted instantly by four pairs of muscles, a reaction far quicker than any possible in the adult oyster. As the larva swam, it extended a slender cylindrical foot out in front of its shell, and it too could be withdrawn into the comparative safety of the tiny shells.

The larva was guided by sense organs important only during its planktonic existence: a pair of dark eyes, each with a light-concentrating lens and a cup of pigmented sensitive cells, and a pair of organs of equilibrium. As members of the drifting plankton, larval oysters not only escaped many filter-feeding predators of the bottom, but were assured of a wide dispersal by making use of the tides. The microscopic swimming oysters were completely ineffectual moving against tidal currents, but they could easily migrate vertically in the shallow water. When the tide ebbed, they sank toward the bottom where they were not likely to be carried out to sea, but rose again when the tide flooded into the bay. Since flood tides generally lasted longer than ebb tides, larval oysters were gradually carried up the bay away from the densest concentrations of adults. Many were caught in the counterclockwise currents and transported over to the eastern part of the bay which, because of its strong river flow, could not otherwise be colonized. Some were swept down in this flow back to the mouth of the bay, where they might find themselves close to their point of origin. Because of this habit of rising into the circulation patterns of a flood tide, the entire bay was offered a liberal sprinkling of young oysters, far more than it could support, and the oysters themselves were retained within the estuarine system.

After two weeks, larval oysters would be ready to settle down permanently. By then their eyes caused them to search actively for darkened regions along the bottom where, foot extended and sampling the texture and composition of the surface, they would creep along. At first the long foot would reach out, take hold, and drag the tiny shelled creature after it, but this locomotion slowed when a suitable surface was found. By stretching its foot far out to serve as a lever, the tiny bivalve would then topple over sideways. Cement from a gland at the base of the foot would affix the shell permanently in place, and the changeover from larval to adult form would commence at once by absorbing foot, velum, and eyespots.

The number of settling oysters, or spat, depended upon many factors related to the nature of the bottom and the region of the estuary. Each larva was an extremely sensitive seeker of conditions. One would always settle on a surface already covered with bacteria and diatoms in preference to a cleaner spot. Soft mud was an almost impossible place of attachment, but hard, stable bottoms composed of rocks and shells, occupied or dead, and washed by reliable currents, would receive a steady rain of young. A large oyster shell, recently cleaned out by a crab, might support two hundred spat at first, but all of these could not possibly survive; that same shell later on would have room enough for only one or two full-grown oysters.

As oysters took hold on the bottom and began to develop extensive beds, the structure of the living reefs which occasionally broke the surface was always at right angles to the current. This was the result of young oysters settling and growing outward where the food-carrying tidal sweep was less interrupted by the existing reef, so submerged bars and reefs ran across the bay, at least in the shallow flats.

Mortality was extremely high among oysters, for all kinds of conditions, catastrophic events, and living enemies made the survival of any one young oyster a doubtful matter. Sudden floods coming down the river and prolonged high temperatures could kill if they persisted beyond the ability of an oyster to remain closed. Sediment was constantly settling over the entire oyster community, the product of bottom-dwelling animals as well as of currents in the river and bay, causing wholesale suffocation. Even if the water was clear above, sediment might be rolled along the bottom, covering everything in its way inches thick with silt. Waste material from the oysters themselves would build up in crevices between them, creating severe difficulties for the youngest and smallest spat.

Disease was common in the oyster population, and had been over the millions of years they had lived in shallow, coastal waters. Different kinds of fungus could attack living tissue or the shell; bacteria and spiral microbes, one-celled sporozoans, flagellates, and roundworms had affected oysters in the past as well as those now living. There was a constant elimination of weakened individuals, resulting in the survival of strong and resistant strains of oyster.

Thousands of oysters were harboring yellow sponges that had pitted their shells and were growing in eroded galleries within the layered shell structure. When the chemical erosion finally reached into the cavity occupied by the oyster, the mollusc quickly deposited more shell material over the hole to seal it off, but not always in time. If the oyster died, the sponge would ramify all through the shell and grow to great size. Similarly, a few types of marine algal growths were able to penetrate the outer layers of the shell, eating out anchorages by means of weak acids dissolved in the water. Small crabs lived inside some oysters where they damaged the delicate gills. Female crabs entered as minute planktonic forms and grew to the size of a pea, trapped in their hosts for the rest of their lives, but the males entered and left at will, seeking out females for mating.

Then there were the predators, small or large, that actively sought oysters for food. The most abundant of these was a small yellowish snail, a drill, which in the lower bay decimated the crop of spat and young oysters. Although it preferred barnacles, oysters were more common, and with a keen chemical sense it found them before their shells were too thick to penetrate. A drill snail rasped at the shell of a young oyster with a flexible, filelike tongue, then placed on the spot a blunt organ that secreted a fluid which dissolved the structure holding the calcium compounds together. This boring organ was held in place for an hour, sealed by the broad foot of the snail, then was removed while the tongue went to work again for a few minutes. As the process continued, a slightly beveled hole grew downward, the drill testing its depth frequently with its proboscis. When penetration was finally effected, the snail thrust its proboscis into the shell cavity and began feeding on the defenseless oyster.

Drills and their urn-shaped egg cases were everywhere in the lower bay, but grew fewer and finally disappeared as the sea water was diluted by half. Above this fluctuating line in mid-bay, oysters were free of their attacks. Another predator, the starfish, was limited in the same way. As a marine creature, it could not tolerate much dilution of sea water, so it was a serious predator only in the lower bay, where, by force alone, it opened the shells of even large oysters. To do so, it applied hundreds of tube feet equipped with suction discs to the two shells, then exerted not only muscle pressure, but a hydraulic force not even the powerful muscle of the oyster could withstand for long. Hours or even a day later, the oyster, exhausted and deprived of oxygen, would relax. Once the shell was open, even a minute fraction of an inch, the starfish would evert its stomach, push it between the mollusc's shells, and digest the fleshy body in place. The stomach, when inserted, usually ensured an even wider gaping due to a narcotic effect.

Starfish in the bay were not great travelers by their own

slow movements, but they, like larval oysters, made use of currents in the bay. Caught in a strong flood tide, one might relax its grip on the bottom, curl its arms upward, and be swept away. At times currents could carry one into regions where it would be unable to survive, but there were always more coming into the bay from the sea outside the capes.

Some predators were small and insidious. Two different kinds of flatworms could creep into an oyster undetected and weaken it until it died. Others used brute force to make a meal of the bivalve. A huge marine snail, the knobbed whelk, used the sharp spire of its shell to break off the edge of an oyster shell, inserted the spire as a wedge, forced the shells apart and devoured the meat with its massive toothed tongue. Blue crabs and rock crabs were often powerful enough to chip away at the edge of a small oyster and get at the animal within. Certain fishes, such as the black drum, would break off an oyster and swallow it whole, crushing its shells between powerful teeth located in its throat, then spitting out a rain of shell fragments that fell to the bottom.

All of these events befell the oyster, yet it remained populous through the bay and provided a habitat for an enormous assemblage of plants and animals. Wherever there were oysters, organisms grew upon them or lived in the crevices formed by their angles of attachment. This bay-bottom world created by the oyster was the truest of all estuarine communities: any one small chunk of cemented oysters was in fact a microcosm of the whole great bay.

On many of the oysters there were clusters of dark, squat sea squirts, each possessing a pair of siphons which drew in and exhaled water for its food and oxygen content. Though they were externally nondescript, sitting there fixed for life, their larval stages disclosed they were members of the same major group of animals that had given rise to fishes and mammals, reptiles and birds. In the distant past, a small creature like the present larval sea squirt, or tunicate, had developed the ability to reproduce while still immature, thus escaping its undistinguished developmental fate of becoming only a grown-up sea squirt. This larva, with a faint suggestion of an internal stiffening and a main nerve cord along its back, and with gill slits in its neck, had been generalized enough to develop over the ages into entirely new kinds of creatures with backbones—the vertebrates. But here, fixed to the bay bottom, were those other descendants, the traditionalists which had remained what their hereditary patterns had told them they must be. Sea squirts were plentiful now, but they were seasonal and would die in the fall, to be replaced by their tadpolelike larval forms in the spring.

Other oysters served as pedestals for spiral chains of slipper shell snails, each clinging to the next in pickaback fashion. Those on the bottom of the pile, the oldest, were females; the ones on the top and the youngest, were males. In between, the snails were hermaphrodites, capable of both male and female functions. Slipper shells hardly moved at all, except to elevate themselves slightly to allow ventilation, but there were other snails in the community that were very active. One, a dog whelk with a cross-striated shell, crept along quite rapidly on a broad, muscular foot, its long proboscis extended and waving about sampling the water. Like the drill, this too was a predatory snail, but one that fed on animals other than oysters. Occasional orange, white-spotted snails with porcelainlike shells browsed on accumulated organic material—spots of color in an otherwise drab community.

Perhaps the most graceful of all the bottom-dwellers were slender, white sea anemones that stretched into the moving layers of water overhead, their fine tentacles spread in a delicate, streaming net. When a drifting member of the plankton was carried into the tentacles, it was quickly paralyzed by explosive stinging cells and eaten. The sea anemones were so translucent that longitudinal membranes were visible within their bodies, as were dark masses of accumulated food. Slow, rhythmic contractions pulsated the length of their tubular trunks, and often an anemone would contract entirely into a globular lump.

Worms lived freely in the spaces between oysters, or else they resided in tubes of mud or lime. The largest worms were the same sandworms that had been present higher in the estuary but here, in a more satisfactory environment, they were several times the size of the ones in less salty water. Although they could retreat into temporary tubes built into crevices, the sandworms spent much of their time crawling through the maze of oysters, their flexible and mobile bodies following every depression and elevation. A sandworm normally sought organic remains for its food, but when the opportunity presented itself, it could evert a muscular pharynx equipped with two amber, scimitar-shaped jaws to grasp living prey.

The most common worm, a much smaller one, had emerging from its head two long filaments that constantly flicked about in search of food and suspended sediment, items that were picked out of the water by sticky mucus, then passed by beating hairs down the length of the tentacles. Food was separated and swallowed, while silt particles were deposited around the body to form a mud tube. The tube was open at both ends; after working on one end for a while, the worm would double back upon itself, still inside the tube, and renew construction at the other end. The result of all this activity by as many as four hundred worms to a square inch was an extensive and possibly hazardous accumulation of mud on top of, and between, oysters. When sunlight filtered to the bottom and illuminated the oyster community, thousands of these translucent, writhing tentacles caught the light and made the whole agitated surface appear to flicker as with a pale flame.

Occasionally these same worms developed a more intimate relationship with oysters by living inside the shells where their U-shaped tubes were covered by pearly secretions of the host, until a blisterlike effect was created. The worm not only extended tentacles into the oyster's cavity where food-laden currents were strong, but released planktonic larval forms so small they were carried out in the exhalations of the mollusc.

Another type of worm grew in profusion where the oyster bed was thick with old shells. This one constructed elevated, serpentine tubes of a limy secretion, securely anchored to a solid surface. The whole mass, a foot or more in diameter, rose high off the bottom. Nearly every tube was occupied by an orange worm restricted to this sedentary existence for its entire life. Sprouting from the worm's head was a rosette of colorful featherlike tentacles, some red, some blue-black, and others yellow or mottled. The tentacles were heavily ciliated and drew water down toward a centrally placed mouth from which food particles were removed. Each tentacle also had a series of small black dots running along the midrib, simple but effective eyes. Whenever a fish or swimming crab passed overhead, the worm retreated down its tube in a flash, plugging the aperture with one highly modified, club-shaped tentacle.

A few of the limy worm tubes were lined with sand deposits, the yellow-brown of the particles contrasting with the chalky white of the tube itself. Poking out from such a narrowed aperture were two pairs of red and white banded antennae; soon their owner emerged a little, revealing itself as an elongated shrimp that busily swept the area clear of organic material. Most of the worm tubes were also covered with a fine, white mosaic pattern resembling miniature cobblestones. Each unit was in effect a little capsule with an opening at one end surrounded by fine teeth. Inside the capsule lived a tiny, delicate moss animal, a bryozoan, which filtered from the water the finest of suspended plankton, mostly bacterial in size.

Other kinds of bryozoans lived on every available surface. Some were fleshy and erect, their blunt pink branches
spreading like fingers; quite different were those that consisted of long and slender strands, composed of a dark
hornlike material. Even more abundantly, delicate bushlike
hydroid colonies rose swaying in the currents. Hydroids,
marine relatives of the fresh-water hydra in headwater
streams and ox-bow ponds, grew in a variety of branching
patterns, each tiny flowerlike individual encased in a cup of
transparent skeletal material. Toward the bottom and close
to the stem of most hydroid colonies, stout, elongated
chambers revealed near-microscopic jellyfish budding from
a central core. There was an opening at the end of each of
these chambers, so when their development was completed,
the tiny pulsating medusa could leave to join the plankton.
They never grew to great size, but matured sexually very
quickly, releasing sperm and eggs into the water. The resulting larva, a ciliated creature of great simplicity, would
finally swim to the bottom, attach, and begin budding the
flowerlike, tentacled individuals that composed the mature
hydroid colony.

Hydroids attracted creatures that might otherwise be
absent or scarce in the oyster community. Often their stems
were places of support for skeleton shrimp which clung by
means of short legs equipped with scimitar claws. These
legs were all at the rear end of a skeleton shrimp's body; the
body itself was attenuated and equipped with two pairs of
flaps which in the female were enlarged and served as a
brood pouch for young. Each skeleton shrimp stood out
from the hydroid stems, bowing and waving its long body,
while powerful claws extended in readiness at the head end.
When a lesser creature happened by, the claws shot out,
jackknifed around it, and brought it back to the jaws which
quickly shredded and devoured the victim.

A deliberate creature that appeared to be all legs made its

slow way through the tangled mass of hydroids. It had no abdomen at all, only a short stiff body with a blunt tubular snout. But its eight jointed legs were long and well developed, the segments next to its body greatly inflated. Organs normally found in an abdomen were packed into the legs where eggs could be seen in the process of formation through the transparent body skeleton. It was a sea spider, but in no way related to the spiders of the land. Together with the skeleton shrimp and much smaller copepod crustaceans of a crawling sort, the sea spiders failed to elicit any response from the hydroid tentacles heavily equipped with stinging cells.

Sea slugs also hovered about the hydroids, feeding upon the tentacled individuals which found no security against these much larger animals by contracting into their glassy cups. The shell-less sea slugs were yellow and brown, handsomely mottled with white, a forest of branching white-tipped filaments on their backs. In part, the filaments served as gills, but were also important to the animal's defense. When a sea slug devoured a hydroid polyp, it ate its stinging cells with no apparent effect. Later, the explosive stinging capsules, separated from their parent cells, migrated to the sea slug's filaments where they became embedded at the surface with their triggers pointing outward, captured weapons ready for defense.

On a still smaller scale, the oysters were covered by layers of diatoms and vast numbers of single-celled animals. The branching, treelike protozoans so abundant at the head of the bay now were giving way to a knobbed, stalked solitary kind that radiated tubular sucking tentacles from its globular body. When a lesser protozoan came in contact with the tentacles, it was quickly immobilized and its cellular contents sucked through the tentacles until it resembled a collapsed bag. Much larger ciliated protozoans lived in tubes cemented lengthwise to an oyster shell, the open end slightly elevated. The animals within thrust out two ciliated lappets which drew in a constant supply of bacteria and organic silt.

Over the fixed community came the mobile creatures, large and small, often more active than snails and the numerous sluggish, stout-bodied mud crabs. Spider crabs, with legs that spread over a foot from toe to toe, ambled across the rough bottom, picking at weakened and dead organisms. Because they shed their outer skeletons less frequently as they grew older, large spider crabs often had small seaweeds, sponges, and moss animals living on their backs. Stocky, powerful rock crabs edged their way beneath overhanging clusters of oysters, their dark-red bodies indistinct in the shadows. Hermit crabs scuttled about, dragging their snail-shell houses with them, into which they could retreat if danger threatened. One large hermit crab, occupying a moon snail shell, was hardly distinguishable from the heavily settled oyster bed surface, for its shell was completely covered with attached, living organisms that included two kinds of slipper shell snail, worms in limy tubes, bryozoans, and a single large barnacle. Inside the shell, flattened enough not to crowd out the hermit crab, were several dozen smaller slipper shells, and a heavily protected scale worm which was able to move about without irritating the crab. Another lesser kind of hermit crab, ducking into its house as the large one strode by, occupied a shell completely covered by a pinkish hydroid colony, composed of distinct types of polyps, each specialized for feeding, defense, or reproduction.

The hermit crabs were a quarrelsome lot; when two of the same kind and of similar size came up against one another, there was an immediate contest of grappling, tugging, and attempts to extricate the weaker from its shell. Shells did have to be changed periodically, for a hermit crab would molt and increase in size, finally outgrowing the old shell it had carried about so long. An unoccupied snail shell was a rare item on the bay bottom, although the largest conch or whelk shells were too big for a hermit crab; if they did not become filled with mud and sand, they served as temporary retreats for mud crabs and small fishes, such as gobies and striped blennies.

A channeled whelk, only a little smaller than the giant knobbed whelk that had fed upon oysters, was moving through a slight depression with its large foot extended far beyond its body. Hovering nearby was a small fish, a caesar, marked by a conspicuous dark stripe running the length of its body, even through the middle of its eye. The fish stayed very close to the whelk, resting against it and sidling about from one side of the shell to the other. The whelk slowed and stopped its progress through the muddy sand, withdrew its broad foot a little, and raised and twisted its shell. From the aperture issued forth the first of several dozen thick disks, all linked together on one side by a tough cord. Each was an egg capsule containing a dozen or more developing eggs embedded in a protective albumen substance. The firm chain of leathery cases kept being produced slowly, settling to the bottom in a long coiling spiral. In each of the fifty-to-sixty flattened capsules development was already taking place, until after two weeks had passed hundreds of tiny whelks, only an eighth of an inch long, would emerge to take up a precarious existence on the bottom.

Toadfish, heavy and warty with a huge gaping mouth, were among the largest predators of the oyster community, although they had little to do with the oysters themselves. Instead, they fed abundantly on crustaceans and, to a lesser degree, upon oyster drills and small fishes. Because some of the crustaceans and the drills were harmful to young oysters, the toadfish had a direct effect upon the success of the oyster populations; without them—and there were many toadfish lurking in the area—there surely would be fewer oysters.

The heavily populated oyster community, with no space left unoccupied by plant or animal, stretched mile after mile across the bay, absent only from those regions where silt had built up to impossible depths, or where the current was so strong that no attachment was possible. Yet this was only one part of the bay's productivity: its plankton populations were enormous; shrimps, swimming crabs, and fishes teemed in open water; soft bottoms were packed with worms, molluscs, and creatures of lesser size. Bacterial populations were astronomical. The intensity of life present in the bay had never been equaled in the river above and would diminish again past the capes at the entrance to the bay, for this phenomenon was a reflection of the warmth of the great shallow basin which, with an unending supply of nutrients, not only supported its own but attracted countless animals from coastal waters to enter, feed, and reproduce. In this midsummer season, the bay was both haven and nursery.

Over a thousand miles to the south, where the huge circulating current of the ocean came in close to semitropical islands with their shallow reefs, fishes were sometimes trapped in the massive flow and carried north. As the chill of northern latitudes began to have an effect upon the stream and its inhabitants, many left and sought coastal waters warmed by the summer sun. The lower bay was a temporary refuge for a scattering of these misplaced fishes. The unlikely appearance of triggerfish, filefish, surgeon fish, butterfly fish, angelfish, and others occurred from time to time near shore. A week earlier, a huge manta ray, fifteen feet from tip to tip of its winglike fins, had come in, seeking food among the rich plankton. But for these fishes it was a transitory existence. Now that the summer solstice was past, the days in which they could survive were growing fewer; lowered temperatures of autumn would exceed their individual abilities to tolerate cold and they would die. There was no way back to the warm waters of the south.

For endemic creatures, the temporary security of the bay provided ideal nursery conditions, and its warmth accelerated the development of juvenile stages unable to regulate their own body temperatures. Eggs and larval forms of animals were everywhere and motile spores of seaweeds often filled the water. Menhaden, a small oily herring, entered the bay in schools numbering hundreds of thousands of individuals covering many acres, and there were hundreds of schools. Their presence was noticeable from the surface, for these plankton feeders spent much of their time close to the top of the bay, gulping in multitudes of diatoms, larval animals, and solitary living cells. When a school rose to feed, it had a stilling effect upon the small waves, while the water turned a faint pink from oils they exuded. From a distance the surface appeared agitated, as though a fretful local wind blew directly down upon it, but when the school drew near, the silvery bodies of the menhaden caught the sun and produced a wide, bright twinkling just beneath the water. They were here for the summer, passing into tidal creeks in bordering marshes to feed.

Not only fishes entered the bay in the warm months of the year, but crustaceans and their relatives as well. A few, at the northernmost end of their range along the coast, sought the warmth of the bay as did the semitropical fishes. A sprinkling of very large shrimp occurred on shallow sand flats, the same animal that swarmed in huge droves in gulf waters over a thousand miles to the south. The most ubiquitous large crustacean in the bay was the active, pugnacious blue crab that dwelt everywhere in the broad basin, and penetrated far up the estuary to explore tidal creeks almost to their headwaters. It achieved a wide dispersal because it swam vigorously with a pair of jointed, sculling paddles, the last of its four pairs of legs. With its other legs, it walked easily across the bottom, digging about in the muddy sand for recently dead food. It was primarily a scavenger, but when the opportunity arose, it could turn predatory and dig out burrowing worms or catch a slow fish in mid-water. There were times when a blue crab might even turn cannibalistic if it found another of its kind soft and defenseless while molting the outer skeleton. Its powerful, sharp claws could not only crack open small molluscs and shred large, dead fish, but were formidable weapons when it was bothered by other crabs or attacked by a much larger predator. If the enemy was too large to be pinched and hurt, the blue crab would use the sharply pointed fingers at the end of its claws as lances to stab into the flesh of an attacker.

Many of the blue crabs were females carrying bulky yellow masses of eggs—perhaps two million of them—cemented to their broad, folded-under abdomens. Whenever one rested quietly on the bottom, she fanned her abdomen up and down to ventilate the developing young, assuring them sufficient oxygen for their rapid rate of development.

Some of the females had already lost their egg masses and were on their way into deeper water, the young having escaped into the plankton as tiny, long-spined creatures totally unlike their parents. During their drifting life in open water, planktonic crabs would go through a number of molts, each succeeding stage differing a little from the preceding one, until the larva resembled a shrimp in form, with a long straight abdomen. Then a major change took place: proportions altered as the body of the young crab broadened and the abdomen, formerly so prominent, now tucked under the body and failed to develop at an equal rate, finally becoming a pointed, insignificant sliver in the male, and a blunted flap in the female.

Farther down the bay, the oyster flats gave way to wide stretches of muddy sand populated by burrowing worms and clams. Small grassy islands rose from the shallows, some of them with thick cathedral-like stands of very tall and straight loblolly pines, their needle foliage bunched at the top. Dead trees and bleached pine stumps extended into the water from several of the islands, providing perches heavily used by herons and gulls between their fishing flights. On islands employed as heron rookeries, a number of trees were dying from the effects of concentrated droppings from generations of birds, wastes which had altered the chemistry of the sandy soil beneath the nests. Tassel-topped reed grass surrounded every island and lined the shores of the bay, often standing well out in the water during high tide.

It was in this region that several great helmet shapes moved in cumbersome fashion across the shallow bay bottom, each trailing a long spikelike tail. Their means of progress was indistinct, although the sand retained impressions of pointed feet and splayed, pusher legs that were hidden beneath the streamlined, hemispherical body. The horseshoe crab, lumbering out of the past virtually unchanged over a span of one hundred and seventy-five million years, was unique in this ocean; the only other related forms were three species on the far side of the world. It was not a crab at all, but distantly linked to land-dwelling spiders. The smooth, rounded outer skeleton covering the body lacked limy salts that strengthened the bodies of crabs and lobsters, so was slightly flexible, but extremely tough. An adult horseshoe crab had practically no enemies in the sea. It passed its slow life in bays and coastal waters, reaching into soft bottoms to devour young shellfish, worms, and other sedentary creatures. Nowhere in the entire river was there anything so ponderous, so archaic.

The largest of the horseshoe crabs now passing over the sand flats had not molted its outer skeleton for many months, and was carrying an assemblage of bryozoans, hydroids, slipper shell snails, and barnacles on its back. Underneath and toward the rear of its body, where the long spine of a tail was jointed, a series of wide, stiff gills slowly fanned back and forth. Several of the gills bore small white flatworms which lived there doing no harm to their host but deriving benefit from the continuous gentle flow of water.

One of the horseshoe crabs began digging into the sand, from which it extricated a small razor clam with its smooth, pointed clawed feet. The creature lacked jaws of any kind, but placed the clam between its five pairs of spiny shoulder joints and crushed it into fine particles which sifted downward, although not before the animal had a chance to suck in fluids and soft body parts.

The horseshoe crab hunted primarily by a sense of touch when it probed the bottom, but it had large, oval compound eyes pointing outward from the sides of its helmet-shaped body, and another small pair of simple eyes on either side of a blunt spine toward the front of its shield. The compound eyes were of special importance to the travels of the animal, for they detected polarized light from the sun, forced into parallel waves as it entered the water. The animal oriented itself by these strongly directional rays of sunlight.

The bay was filled with horseshoe crabs, now widely dispersed, but only two months earlier they had swarmed onto the protected beaches by the hundreds of thousands, riding up on high tides around the time of the full moon. Each evening for a week or so, the shallow water had poured forth massive invasions of the animals, mostly coupled together with the male mounted above and toward the back of the female. They had made their slow way up the beach beyond the small waves where the female had laboriously excavated a pit and spawned long ropy strings of pale green eggs.

During the following days, the beach had been greenish with millions of eggs uncovered prematurely. Most eggs, however, developed within the moist sand to the point where the tough, opaque green coat burst open, releasing a transparent egg sac which quickly inflated, revealing a tiny, tailless horseshoe crab within. The embryo waved delicate legs and gyrated slowly within its bubblelike prison. When tides grew high again and washed over the area where the eggs lay, sharp sand grains cut the egg membrane, and the pale little animal was freed into the vastness of the bay where it did not have very much chance of surviving. During the time the eggs had been in the sand, both shore birds and land birds had fed upon them; released into the water, larval horseshoe crabs became victims of almost every bay predator, especially killifish that crowded into the shallows. The safety of the race lay in its fecundity; more than enough young horseshoe crabs survived to maintain the populous role of this relict from the distant past.

Where was the river now? It was very much present, despite being dispersed throughout the wide bay. Part of it still flowed close to the surface along the western shore of the drowned valley. More importantly, it moderated the nature of the entire bay by dissolving the invading salt water of the sea, establishing conditions that were unique and of the greatest importance to resident populations. The lower bay was more stable than the upper region near the mouth of the river estuary had been, with a lessened tidal rise and not so wide a variance in salt content. Because the environment more closely resembled the sea, oysters were no longer safe from the attacks of drills and starfish. Relatives of the starfish, even more marine in their requirements, were able to penetrate ten or fifteen miles up the bay: brittle stars with slender, snakelike arms huddled under shells and rocks, sea cucumbers burrowed into loose silty sand close to shore, and spiny sea urchins wedged between rocks that had crumbled from the bordering cliffs. Plankton populations now included predatory arrow worms as glassy invaders from the open sea, and swarms of soft comb jellies, their eight rows of pulsating ciliated combs glistening in the sun with refracted rainbow colors.

Comb jellies were so densely packed in some areas, they calmed the surface and almost completely eliminated lesser plankton by their feeding.

The water of the lower bay was filled with sound. Toadfish tucked away in crevices uttered solitary, hollow grunts, while sea robins, walking along on their clawlike fins, chirped loudly back and forth. Near the edge of an oyster reef, a mid-water din increased in intensity as a school of several hundred croakers swam slowly overhead, drowning out every other sound by their collective roar. When the fishes had passed and their guttural cries receded into an indistinct background of sound, sea-voices in all their variety were heard again, each species identifying itself to others of its kind. There were peeps and squeaks, whistles, deep drumming noises, harsh cries, snaps, and mechanical stridulations.

The cliffs containing the lower bay were almost vertical and rose nearly one hundred feet above the water. They were composed of five or six distinct strata of sand, rocks, and gray clay, the uppermost layer heavily vegetated and gently rolling. Vertical dark bands of ground water seeping from several of the strata gave the cliffs a checkered look, except where huge scalelike chunks of earth had fallen to the narrow beach below. High in one layer, cliff swallows flew in and out of their nest holes that pocked an area fifty yards long.

One of the lowest layers, only five feet above the beach, was thirty million years old and almost white from the enormous numbers of fossils that had been exposed as the cliff wore away. It revealed a complete community from ages past; the only missing clues were those to the vegetation of that ancient shallow sea. The animals were still there, their remains crumbling out of the soft rock in an unending rain of fossils that fell to the beach and were washed out into the shallow water where they littered the bottom. Some were minute, such as the limy shells of one-celled foraminiferans and of little crustaceans that had lived in hinged, clamlike shells. Extensive colonies of flat moss animals still clearly covered many shells, which also bore signs of hydroid attachment. Prehistoric oyster shells had precisely the same signs of damage by boring sponges as their descendants living only a hundred yards away and identical distinctive holes of some ancient predatory snail. There were corals present, a sure indication of an earlier warmer climate, masses of worm tubes, and molluscan shells of every shape and size. Corkscrew shells, crab claws, heavy ribbed snails, blood clams, and venus clams lay clearly exposed, but the most common fossils of all were giant scallops, some of which were still hinged together and bore clusters of huge barnacles on their upper shells. There was even evidence of more highly developed life in the form of bones and teeth of rays and sharks, some of the last so large that they must have dropped from the mouths of great predatory fishes over forty feet long.

The fossils lying scattered beneath the waters of the bay again entered the world of the living after a sterile isolation of thirty million years. They provided area for the settlement of modern oysters, bryozoans, sea squirts, and all the other bay creatures that sought space for attachment. Eggs were laid upon them, and small crabs and worms sought shelter beneath them. Every available new surface supported a succession of colonizers in quite definite patterns. Within hours after a fossil shell tumbled into the bay, it was covered with bacteria and soon after, by diatoms. Protozoans and hydroids were the first animals to settle, but barnacles and bryozoan moss animals were not far behind. If the shell remained close to shore, small seaweeds sprouted, perhaps growing long enough to catch the current and carry the shell away. Chance and the environment would determine what the final climax population would be, but sponges, sea squirts, and mussels commonly grew into so dense a cover that no other animals could succeed them.

Large fallen rocks along the shore also presented opportunities to plants and animals seeking intertidal conditions. These organisms, which could survive neither completely submerged nor completely exposed, had not lived successfully farther up the bay because of their marine affinities. Here, in saline waters that rhythmically rose and fell, protected from the heavy surf outside the bay, intertidal life thrived.

A strong ebb tide was running. As the full moon was just past, the water now fell close to its lowest point in the month, exposing multitudes of organisms on all available rock surfaces. Each form of life occupied a distinct horizontal zone according to its ability to withstand exposure to the air. During long years of slow development, intertidal plants and animals had found surcease from competition by moving up into areas and conditions that other marine life could not tolerate. A few had been able to move very far indeed, but most were restricted by their inherited limitations to regions inundated at least half of every tidal cycle.

The crests of rocks standing at least a dozen feet out of water were capped by an orange lichen which grew securely attached to the rough surface. The lichen, an association of fungus and alga, erected little shallow cups within which spores were produced. Below the orange lichens and a few other flattened gray-green ones, the brown rock was bare for several feet until, quite abruptly, it turned black from partly dried blue-green algal mats moistened only at high tide. The perfectly horizontal black zone was followed immediately by a scattering of small, gray barnacles, the most resistant creatures from the sea. They quickly gave way to another larger barnacle, less tolerant of exposure but far more numerous, that created a broad, white band reaching down below the mid-tide level.

Each barnacle, hidden within a cone composed of six interlocking plates, was now sealed against moisture loss at low tide by a pair of double valvelike gates occupying the aperture of the cone. When waves and the tide covered it twice a day for several hours, the valves would open and a barnacle would begin rhythmically to sweep the water for plankton with long feathery legs. From a strictly anatomical point of view, it was attached upside down to the rock.

Within the shelled cone, the complex anatomy of the living barnacle did not clearly reveal its crustacean affinities, but as a larva, there was no mistaking it. The tiny ellipsoidal larval forms were members of the plankton and so numerous that when a reasonably clear surface was available they settled quickly, after smaller organisms had pioneered the way by covering the area with a bacterial film, diatoms, and other microscopic life. When a barnacle larva found such a surface, it left its free-swimming way of existence and began exploring the area for the right place to attach, testing the surface by means of highly sensitive antennae equipped with flattened, chemically responsive hairs. It had quite definite preferences: depressions rather than flat areas; shallow pits over deep ones; elongated depressions over rounded ones; wave velocities of about three miles an hour; and attachment no closer than an eighth of an inch to another barnacle. Under such conditions, a heavy set was assured. When it found the right place to spend the rest of its life, it turned over on its back, cemented itself to the surface, and began constructing the limy conical shelter which was part of its outer skeleton. The overlapping plates of the cone allowed for growth as the barnacle increased in height and diameter, radiating out from the circular base of attachment. Muscles in the otherwise soft animal not only operated its legs and valves, but secured the side plates to the base, so there was little danger of accidental removal, especially by waves. When one barnacle came in contact with another, a slow contest began:

by growth and minute inching movements, the base edge of one cone would attempt to undercut the other. Where barnacles settled in dense populations and were crowded together, outward growth was impossible and they grew upward as elongated tubes, for the highest barnacles had the most effective sweep in the plankton-laden water. But growing tall was hazardous; where a small spot of rock had been cleared by predators or accidental means, the tubular barnacles were easily undercut by younger ones and were susceptible to tidal currents and waves. A single conical barnacle was almost perfectly suited to intertidal life, yet its colonies were not stable and often tended to be in jeopardy.

In tightly packed barnacle communities there were dead ones, killed by attacks of dog whelks and oyster drills, or destroyed by some other means. With the demise of an individual barnacle, its body and hinged valves soon disappeared, leaving a hollow cone still securely cemented to its neighbors. When submerged by the tide, many of these empty barnacles were occupied by small isopod crustaceans. They could be seen creeping over the surface searching, or with heads buried in empty barnacle shells hunting for decaying matter, small worms, and other tidbits. Often they swam out six or seven inches from the rocks into the swirling water and, with a graceful loop and roll, circled back to make a successful landing on some other portion of the colony.

The surfaces to which barnacles attached were not all flat and facing the sea. Where crevices served as sluices, waves foamed far back into the rocks. Barnacles, snails, and other lower intertidal creatures, taking advantage of such a spot, could live much higher above sea level than on exposed surfaces. At the upper end of one such crevice, a ruddy turnstone stood watchfully with one foot raised, then descended to lower, damper regions where it probed with its stout bill for isopods, worms, and whatever else it could find. As it descended the depression, it often popped its head above the rim to take note of the surroundings.

Beyond the white zone of the barnacles lay the domain of seaweeds—brown rockweeds first of all, with their tape-like strands and ovoid bladders draped down across the rock. When the tide rose and covered them, the floats held tough, flattened ribbons erect as they swayed back and forth in the rolling waves. Because of a slippery, mucus coat, these brown seaweeds failed to rub abrasively against either the rock or one another, so injury was almost nonexistent. The physical and chemical properties of the rockweeds were such that practically no other organisms could attach, plant or animal.

Where the rockweeds grew fewer near low-tide level, broad flat sheets of sea lettuce and dense clusters of blue mussels formed the basis of the next congested community. The mussels were not rigidly fixed as were oysters, but anchored by tough threads, several dozen to each mussel. Whenever tides surged in, the mussel bed was jostled but without harm. If a mussel became detached and was fortunate enough not to be swept away, it would at once secrete more threads to refasten itself. Mussels were quite indiscriminate in their choice of substrate on the rock surface, attaching to one another as they did to the rock, to logs wedged in crevices, or to any other firm surface. The whole area was covered with small white dots that had been the cemented ends to mussel anchor lines, now broken and gone.

Moisture and shelter from the hot sun, available beneath the clusters of mussels and beds of sea lettuce, were conditions sought by newly arrived young mussels and an array of periwinkles, worms, and crustaceans. At times of heavy larval settling, there would be hundreds of young mussels to a square inch, the same area that would be occupied by only two or three adults later on, so the mortality rate assuredly was very high. Flatworms, roundworms, and small segmented worms worked their way through the dense forest of mussel anchor threads finding moisture, food, and quiet from even the strongest currents and waves pounding overhead. They were joined by several kinds of crawling, elongated crustaceans, while skeleton shrimp clung to the threads bobbing about in their search for food. This was a good region for predatory snails to deposit clusters of yellow urn-shaped egg cases, and for their young to begin life. Limpets browsed upon algal coatings of the rock surface, returning always to the same spot when the tide fell so their conical shells could effect a tight seal. A bright green scud, very similar in shape to crustaceans in the ox-bow pond, lived under the fronds of sea lettuce where its color made it almost invisible. The greenness was derived from chlorophyll in microscopic algal cells contained beneath its transparent outer skeleton.

Sea lettuce continued briefly below the lowest tide level, but soon gave way to several different brown seaweeds, such as sea colander and small kelp. Still lower, delicate red seaweeds were so fine and wispy as to make their presence in the swell of waves seem precarious. Large and sturdy sea anemones, hundreds of short pink tentacles displayed at the end of a thick, brown body, stretched into the rolling water to trap plankton and an occasional errant shrimp or worm. Little goggle-eyed blennies, fishes imbued with strong territorial instincts, poked their heads out from retreats ready to capture any smaller creature that happened to come by. Here too were starfish, sea urchins, and a host of marine animals associated with the shore, yet unable to withstand exposure to the air.

The marine zonation from sea to the highest reaches of the tides was mirrored by an invasion of the tidal zone by creatures of the land: larval midges built tough little tubular cases directly exposed to waves and tides. The elongated wormlike larva in each tube had special salt-secreting fila-

ments at the end of its body which made it possible for the
insect to tolerate conditions normally lethal to insects.
Adult midges were everywhere on the rocks now that the
tide was out, limited in their flying, but walking briskly
across rocks, seaweeds, and mussels seeking places to lay
eggs.

An osprey appeared high overhead sailing on updraft
currents created by winds flowing across the bay and deflected by the cliffs. The fish hawk held its wings rigid,
curved back a bit as it rose upon a column of air, and
skimmed straight out across the water, head bent down as it
searched for food. It turned into the wind and flew with
heavy continuous strokes before soaring once more. Suddenly it hovered in one spot, wings beating rapidly, then
plunged toward the water, holding its wings in a half-
closed position. The large bird struck the surface with a
resounding splash, sending sheets of water sparkling into
the breeze, nearly disappearing from sight but for its wing-
tips. The osprey rose from the surface with difficulty,
shedding spray from its plumage, struggling to stay aloft
with its prize of a two-pound weakfish clutched head foremost in powerful talons. It labored to the shore, never
reaching an altitude of more than a few feet, and came to
rest on a high rock jutting out from the narrow beach.
There the osprey, one talon imbedded in the still-moving
fish, looked down intently at its prize. Soon the fish ceased
its struggles and the bird, now nearly dry, took off and flew
powerfully along the face of the cliff which began to dip
toward the level of the bay. The shoreline, protected by
the capes not far away, developed into an extensive salt
marsh, reaching inland for nearly a mile.

The osprey flew straight to its nest in the top of a partly
dead loblolly pine immediately behind the low-lying
marsh. The nest was a remarkable structure, having been
occupied by successive generations of ospreys for nearly
half a century. Each pair had made its contributions to the

structure in the form of sticks, seaweeds, pieces of sod that had grass growing out, cedar bark, pine branches, shells, bright pebbles, and bones. It occupied the whole upper part of the tree, with sticks over ten feet long forming the base of the nest which was itself about eight feet in diameter and five feet high. The floor of the nest was solid with packed sand and refuse, cementing the structure in such a way that not even the strongest winds could blow it apart.

The male osprey braked and hovered over the nest, then descended easily, still carrying the fish with both talons, one held directly in front of the other. Three half-grown young that had been crouching motionless set up a clamoring which was intensified as the female also arrived from a short flight over the marsh. The male picked vigorously at the fish, ripping off the head, but it was his mate who, with wrenching, twisting motions, tore off strips of flesh to feed the young. Now that it was midsummer, the young birds were growing rapidly and it took all their parents' time to feed them. No longer did the female have to crouch over the downy chicks shielding them from the sun's heat, for their camouflaged plumage offered adequate protection at last. Within the next two weeks, the young ones would begin wing-beating exercises while holding tightly to the edge of the nest, an activity that would continue to the time when they would leave the nest for short, precarious, and tiring flights above the tree to nearby perches. Even after they had taken to the air and were adept at braking and landing in a tree, they would return to the nest to be fed; it would be even longer before they were successful in their attempts at fishing. A young osprey would dive repeatedly, clumsily missing fish after fish, before learning the knack of capture and rising into the air with a heavy load to carry.

Ospreys were common summer birds around the marsh-lands, occupying nearly three hundred nests in a twenty-square-mile area. Despite its size and power, the fish hawk was not sovereign, but subject to fish-robbing attacks by bald eagles also living nearby. The osprey's fawn-colored eggs, blotched with rich brown, might be destroyed and eaten by audacious fish crows after the parents had left to hunt. When an osprey flew along the wooded border to the marsh, it was often followed by wheeling, screeching flocks of blackbirds and grackles which invariably succeeded in driving it out of their territory. The osprey was exclusively a fish-eater, but in the vicinity of its nest where defense of young and territory was vital it would attack herons, buzzards, even eagles, if they came too close.

The osprey, through its food habits, linked the river, bay, and marine world with that of the land. The weakfish caught earlier by the male had reached maturity by feeding on a wide variety of bay creatures—the common sand shrimp of shallow water, small mud crabs, sand worms, young razor clams, and algal filaments. The nourishment they had provided, about one-tenth of which had been locked into the substance of the weakfish during its life, was now removed from the bay system and taken inland where it supplied the young ospreys with material they required for growth.

The vast salt marsh that stretched out below the ospreys' nest, appearing monotonously the same with its low-lying grasses and featureless expanse, was interrupted only by a network of meandering tidal creeks. Neither land nor sea, the marsh did not support many different kinds of organisms, but the enormous numbers of those present provided essential enrichment to the bay and coastal waters as decayed material flushed out with every tide. Salt marshland, without question, was the single most important transition from land to sea anywhere along the entire continental coastline.

Three thousand years ago, when the water level was much lower, the region had been an open flooded area within the major confines of the bay. After marsh grasses had begun to grow along the shoreline, the sediment they accumulated stretched outward, keeping pace with the rising bay level. As the marsh expanded, it preserved its major characteristic of absolute flatness. Without a cover of shade trees, it was exposed to air and sun until the extreme tides of each monthly cycle flooded it for a few successive days, or when offshore storms drove water into the bay, piling it up until the winds subsided. Mud and peat, shallow toward the present shoreline but over fifteen feet deep below the ospreys' nest on the ancient shore, preserved a record of all the conditions responsible for marsh formation. Toward the surface, hard black clay and loose silt of the present marsh were interspersed with root fibers to a depth of two feet, followed by older layers of soft clay, peat, sand mixed with clay, and finally, the original bay bottom of hard-packed yellow sand.

The marsh, created by tidal activity and prolonged geologic change, was a world of extremes. Unlike the bay outside, there was no time of slack tide when water stood still in the creeks; as soon as flooding ceased, the flow immediately reversed into a strong ebb tide. In less than a minute, plankton that had been carried upstream were rushing back down again. Most of the tidal creeks draining the marsh were so scoured by the exchange of water that they were deep and swift, although they eventually coalesced into broad and quieter streams leading to the bay.

The curving meander patterns of marsh creeks had remained much the same for centuries because their peat banks were solid and resisted change. A few creek channels were so located that they were prime avenues for flood tides, while others assumed the brunt of drainage when the tide ebbed. The volume of flow into the bay was always greater than an invading high tide, for many small freshwater streams from inland entered the marsh along its back boundaries. Here killifish and muskrats lived; the gray head of a diamondback turtle emerged from the opaque water, its white triangular patches on either side glistening in the sun. The turtle, carried by the tidal current, ducked back almost reluctantly to scull to the bottom, where it half-swam, half-crawled along the muddy sand.

The uniformity of level of marsh mud was largely due to the stilling effects of a dense cover of cordgrass. Water was able to enter during high spring tides each month, but because of the thick vegetation it moved slowly and deposited silt evenly across the entire surface.

The cordgrass was of two kinds: a tall variety that fringed the creeks and small salt ponds, and a smaller form that constituted the bulk of all marsh vegetation. Just behind the stands of tall cordgrass, succulent salt-filled stalks of glasswort appeared, growing through mats of green algal filaments that lay upon the mud surface. Where the marsh began to rise a little farther inland, spike rushes grew in a fringing band, safe from the invasion of salt water except under the most unusual circumstances.

All marsh plants had to be tolerant of excessive salt concentrations. Not only would incoming tides affect them periodically, but muds of the marsh were permeated with salts that had accumulated through evaporation of sea water. Small pools dotting the flat expanse were often so salty no plants could grow in them. Some of the pools were ringed by a heavy crust of salt crystals. Yet after a severe rain these same ponds would be filled with fresh water, placing a different stress upon whatever organisms they did contain. The range of conditions on and beneath the surface was so great that only the most specialized and tolerant plants and animals were able to survive, every one of them possessing highly developed means of resisting sudden and severe changes in temperature, saltiness, and the presence of water. Cordgrass, the dominant plant, absorbed salt through its root system, but was able to get rid of it through tiny pores along the edges of its tall, slender leaves. Acres of marshland glittered in the midday sun from the refractile salt crystals decorating the blades of cordgrass.

Temperature could be a problem restricting animal life on the surface of the marsh. The dark mud absorbed and held intense, penetrating heat from the summer sun, yet six months later the entire marsh community would be covered by crumpled and broken sheets of ice.

Because of the high temperatures and extensive bacterial action that consumed oxygen in marsh mud, the creeks and pools often were very deficient in this vital gas. Dead fish and crabs occasionally floated down the marshland waterways, but what combination of factors caused their demise was not at all apparent. If plant and animal residents were to make the marsh their permanent home, they had to withstand an environment that would be hostile to all but the most highly adapted forms of life. The dynamics of the marsh, fluctuating though they were, nevertheless had a certain stability. Highly evolved marsh organisms and the regular tidal flushing were responsible for making this the most important of all shorelines.

In the river estuary where plant plankton were not very abundant, there had been a gradual increase in surplus chemical nutrients necessary to sustain plant life. Some of these foodstuffs were used later by plant plankton close to the surface of the turbid bay, some escaped into the sea, but most were trapped in the tidal seesaw and at one time or another carried onto bordering marshlands. There they provided the initial growth materials necessary for the success of cordgrass and blankets of mud-dwelling diatoms, in both of which newly formed chemical compounds stored energy trapped from the sun. Cordgrass was the world's most productive, energy-rich land crop, and a square inch of marsh mud contained one hundred times as many algal cells as an equivalent area of surface bay water.

Once an unbroken cover of vegetation was established, the marshes were largely self-sustaining. Because of an overabundance of nutrients derived from plant decay by bacteria and other microbes, the lower bay and nearby coastal waters were supplied with rich chemical nourishment at a steady rate as flood tides washed away marsh muds and particles of decomposed plants. The nutrients were used at once by marine diatoms, which soon were grazed upon by small planktonic animals and they in turn were caught by a series of predators. The result was a complex but nonetheless direct link between large predatory fishes swimming far offshore and the immensely productive community of a salt marsh serving as a vast storehouse of chemical energy, a constant source of wealth being distributed to life in the ocean. The physical presence of the river estuary and the life it supported was having a profound effect upon the sea and its populations.

Extremes in conditions made the marsh a difficult place for animals to live; only a few kinds tolerant of environmental excesses flourished in great numbers. When no birds or muskrats were nearby, and the sun warmed the steep mud banks along creek channels, thousands of fiddler crabs emerged cautiously from their burrows. Then, more boldly, they walked slowly across the mud surface, dimpling it with their pointed feet. Females used both their small claws to convey bits of mud and decaying plant material to their heavy jaws hidden behind a set of complex mouth parts. The males could eat only half as fast, for one of their claws was enormously enlarged and totally useless as a feeding appendage. In fact, it was also nearly useless as a weapon of defense, but its light-yellow color was conspicuous; when it was waved rhythmically back and forth during breeding season, it was noticed by females and other males as a display of courtship and territory. At times, the fiddler crabs would leave their burrows and scuttle through the cordgrass, rustling like a new breeze as they migrated to different feeding grounds or reacted to a sudden movement from outside their mudbank world.

The lower banks of tidal creeks were pitted with the round holes of these small crabs which only seldom intruded upon the territory of a larger, red-jointed fiddler crab living higher and farther back in the marsh. The small fiddlers excavated burrows that went deep into the wet soil and into which they retreated behind barricades of mud when the tide rose. They were so conditioned by inheritance to the pulse of tidal cycles that they would continue to rise and descend in their tunnels, keeping perfect tidal rhythm, even when some unnatural event, such as the sudden isolation of a meandering creek by a collapsed bank, prevented water from reaching them.

An entirely different sort of crab lived behind the crest of the creek bank. The purple marsh crab also excavated tunnels in the mud, but the holes were oval and frequently protected by walls and a mud roof, forming a porch. When water covered the marsh during monthly high spring tides, the crab plugged its burrow solidly with mud, not to keep dry—the whole lower portion was filled with water—but to prevent erosion and possible entry by predators. The upper interconnecting tunnels, running horizontally not far below the surface, were filled with air and usually occupied by several marsh crabs living in harmony.

The population of purple marsh crabs was not so great as that of the yellow-clawed fiddlers, but the half-dozen occupying each square yard above the creek margins performed an especially important role: they were one of the few kinds of small animal to eat living cordgrass. One was busily at work pulling down blades, which it then cut into manageable lengths. Picking one piece at a time, the crab thrust an end into its mouth and devoured the short piece in its entirety. The result of the marsh crab's activity was apparent enough, for near its burrow were wide swaths in which all that remained was a stubble of cordgrass. Waste products, left outside the burrows, were composed of partly digested plant material that washed away, enriching the water with nutrients for both plants and animals. Particles of these wastes, teeming with bacteria, were used directly by filter-feeding animals swarming in the plankton and residing on the bay bottom.

While marsh plants were its primary food, the marsh crab was a predator as well, waiting quietly for a fiddler crab to approach. Then with a quickness and agility unlikely in such a stocky and heavily built creature, the marsh crab darted out, seized the smaller fiddler, and tore it to pieces, devouring all but the toughest parts of its outer skeleton.

Only a few other animals lived habitually on the marsh surface. White marsh periwinkles browsed among the algal mats, and small, dark-brown, air-breathing snails were common on the stalks of cordgrass. In some of the less saline pools, larval wrigglers of a salt-marsh mosquito lived, at times in such concentrations that the shallow water seemed to boil with their activity. When a white-tailed deer began walking into the marsh, dark clouds of adult salt-marsh mosquitoes and vicious greenhead horseflies rose in audible frenzy seeking a blood meal, and the deer fled.

Muskrats were the largest permanent residents of the marsh. Their brown, domed lodges, looming up against the low horizon, were scattered across the innermost portions of the green cordgrass where fresh water diluted the tidal surge. Each lodge was constructed of plant stems and old sticks, plugged and held together with peat and mud. Some lodges were a dozen feet in diameter and rose two feet above the level of the marsh, built on platforms of decayed plant material and mud. Radiating from every lodge was a complex system of runways through the cordgrass and shallow, water-filled canals. Muskrats could venture quite far inland, but were not at home on land and were far safer from a predator's attack in the water or running across the screened vastness of the salt marsh.

An old male lived in one especially large lodge. He was outside using his sharp yellow teeth to cut short lengths of cattails that grew along the land margin under the osprey's nest. This was somewhat unusual, for he generally was more active at night when it was safer to be abroad, but today he was hungry. After gathering a bunch of six-inch lengths of blades, he took them to a bare, smooth place along one of his runways and proceeded to eat. Plant and animal refuse surrounding this area showed it was one of his usual eating spots. He took food back to his lodge only when it was collected in the immediate vicinity, so the interior of the lodge was clean and its damp grass bedding proved it was used mostly as a place of refuge and rest.

When he finished eating, the muskrat slipped silently into a nearby tidal creek and headed for one of his canals. He swam powerfully, using his partly webbed hind feet which were made a little more effective as paddles by rows of stiff hairs extending out from each toe. He swam along the shore, where the current was reduced, with front legs pressed to his body and hind feet alternately folding and splaying out with each stroke. When he came to the canal and swerved into it, his flattened, scaly tail suddenly stiffened and bent, serving as a rudder. Nearing the lodge, he quietly ducked beneath the surface and entered one of two entrances, emerging inside a chamber that was several inches above water level. He shook vigorously, sending a spray against the rough walls of his home, but this was only to remove water from his coarse, glossy outer coat; the fine, dense fur lying beneath prevented his skin from ever becoming wet. With great care, he began grooming himself, using his small clawed front feet as combs which he licked frequently.

Some lodges along the rim of the marsh contained females with five or six growing kits, but many others housed only solitary muskrats who defended their territory with great vigor against others of their kind, uttering warning cries and grating and clicking their teeth in an impressive display. Should an encounter or invasion of a lodge result in combat, a muskrat's long, yellow incisors could inflict serious and fatal wounds. The old male was scarred from past encounters.

Often the lodges were inhabited by animals other than muskrats. Because they were quite distinct from the monotonous marsh and the low-lying grassland behind it, lodges were sought by a variety of insects, spiders, and other creatures. Snakes had burrowed into the loose walls of several mounds which also contained nests for the eggs of spotted turtles. In the winter, when ice in the walls cemented a deserted lodge into an impregnable retreat, it might be used as a secure place of hibernation for different animals. Skunks, which were very common along the bay shore, were especially apt to seek refuge there, not bothered in the least by the odor of the muskrat permeating the nest, the result of oily secretions from glands in the rodent's lower abdomen. Not many of the lodges were available to other animals, however; without constant attention by their builders, they usually wore down and disintegrated in three or four years by the combined actions of high spring tides, strong winds, and hungry predators.

Tidal creeks could offer little in the way of a permanent bottom to burrowing animals, but as waterways they were well populated. The gray, white-backed heads of diamondback turtles poked above the surface frequently, establishing their position as they were carried along in the strong current. Often these bluish turtles climbed upon clods of marsh mud that had fallen from the banks and served as small, temporary islands convenient for basking in the summer sun. The water, warm and filled with bits of decaying matter, supported hordes of killifish, silversides, and anchovies. Young menhaden spent much of their early lives here feeding on the rich plankton. Many of the fishes present were themselves planktonic, having just emerged from floating eggs and still equipped with yolk in bulging bellies. Their mouths had not yet developed, so they were unable to feed and were little more than motile embryos.

Larval crabs swarmed in the turbid water, mostly the earliest, long-spined stages of fiddler, marsh, and mud crabs. Tiny opossum shrimps were the most abundant animals by far, now spending the day hovering close to the bottom of the creeks. Later in the day, they would rise, until during evening hours they would populate surface waters almost to the exclusion of all other creatures. A mysid's daily rise and fall was due primarily to changes in light intensity, but the difficult maneuvers succeeded only because of the presence of organs of balance located toward the tip of its abdomen. The way in which a tiny grain pressed upon nerve endings served to stimulate a mysid to take corrective action while swimming energetically. Swimming was not just a means of locomotion, but the only way in which the animal could feed upon fragments of decomposing material and living plankton, filtering these particles from the water by means of closely spaced bristles on its legs and other appendages.

Where the level of the marsh fell low enough to allow regular flooding twice a day, the cordgrass disappeared and nothing was left but wide mud flats inhabited by enormous numbers of mud snails, migrating in distinct but crowded populations from one feeding site to another. Each mud snail, occupying a battered, black shell, fed upon organic ooze and the thin, invisible blanket of diatoms that stretched over the silty surface. A dead fish stranded on the mud banks was completely covered by the scavenging snails, which had left irregular but purposeful trails across the flats as they plowed their way to this source of food. Segmented beak-thrower worms burrowed beneath the mud, each equipped with circular rows of sharp teeth on a protrusible pharynx. Where the mud flats dipped down enough to be covered by an inch of water, pink and red rosettes spread upon the soft bottom signifying the presence of fat, tube-dwelling worms that stretched long contractile tentacles outward to capture food.

Ribbed mussels, larger coarser relatives of the blue mussel in the bay, were embedded in pockets of more solid marsh sod, from which not even strong tidal currents could flush them out. Under the ribbed mussels lay tightly coiled, slowly writhing masses of red and yellow nemertine worms, each equipped with a long eversible proboscis capable of being extended far in advance of the body to collect food particles trapped in the mud pockets.

Birds were nearly the only regular visitors to the salt marsh. They ranged in size from a tiny long-billed wren that clung to sedges at the rear of the marsh, to egrets and green herons stalking through the wide expanse of cordgrass. One green heron paused briefly on a muddy promontory where two tidal creeks joined. With deliberate steps it approached the water's edge where schools of killifish were attempting to maintain their position in the now-flooding tide. It crouched down, head cocked back slightly, until the fish passed directly beneath; with a sudden lunge, it caught one killifish just behind the head and quickly swallowed it. The heron straightened up, ruffling its crest feathers, and watched for more activity, its wide yellow eyes gleaming in the dark plumage of its head.

The most common of the marsh birds, the clapper rail, was at the same time the most secretive and went largely unseen by predators. Two rails, separated by only a few yards, walked through the dense cordgrass with precise and dainty steps, their short stubby tails flicking. At each step, their heads pumped back and forth, as they searched for living food. With their long, curved bills, they picked one marsh snail after the other from the blades of grass, a favorite and constant supply of nourishment. As they approached an open bank, the birds hesitated, then darted out and caught several small fiddler crabs before they had a chance to duck into their burrows. A diamondback turtle, startled by the birds' appearance, slid with a splash into the turbid creek, causing the rails to turn and flee silently into

the cordgrass, heads held low and stretched out, along
well-traveled avenues of escape. One of the rails crouched
against the mud becoming almost totally invisible, while the
other darted on. From where it disappeared in the dense
grass there was a rapidly repeated call, which was picked
up by one rail after another across the marsh, until the
whole area resounded with the loud clacking cries.

Eventually, many of the tidal creeks joined a larger
marsh stream that entered a wide shallow area partly protected by a sand bar. Tidal velocities diminished here, creating a well-flushed basin which teemed with life. The
bottom was carpeted with delicate green and red seaweeds
that could not withstand heavy wave action, and by sulfur
and redbeard sponges. Hermit crabs and rock crabs foraged
in the quiet water with throngs of shrimps and lesser crustaceans. In a clear area where washed sand lay thick across
the bottom, a summer flounder disclosed its presence only
by a gentle fanning of an exposed gill cover, its two eyes
protruding and swiveling about looking for prey. It had
entered the tidal basin over a year ago as a tiny, normally
shaped juvenile fish; as it grew, it had become asymmetrical, tipping over on its right side. During growth when it
fed heavily upon opossum shrimp, its right eye slowly
migrated across the top of its head coming to rest on the
left side, while its body became broad and flat, heavily
pigmented on the upper side and colorless underneath.
After an unprofitable period of waiting in the sandy basin,
the flounder rose from the bottom, shedding sand as it
swam with vertical undulations to a new spot. With a
sudden, convulsive shiver, it sank again into the sand until
only its head remained exposed, colored so like the bottom
that it was barely distinguishable.

Where the basin grew deeper toward the inlet reaching
into the bay, the bottom supported a heavy stand of eelgrass, its long blades bowing to the flooding tide. As a stalk
of eelgrass emerged from the sand, it was yellow, but with

growth it soon became green. A foot above the sand, each blade of eelgrass supported an indistinct coating that became increasingly heavy and darkened the blade farther out. Initially the coating was composed of masses of sedentary diatoms and tiny, erect algal filaments that branched into pointed Y's. Halfway along a blade of eelgrass, hydroids began appearing, then white spiral worm tubes and clusters of starlike sea squirts. Tiny free-living worms and elongated copepods dwelt in miniature forests which grew from each side of a blade of eelgrass. Larger crustaceans, relatives of the pond scud and green amphipod, browsed through the loose diatom cover, ducking down into the overgrowth to escape the current and the nibbling of predatory killifish. Each blade supported a complete association of nearly microscopic plants and animals, distinctly graded in complexity from the base out to the frayed tip two or three feet away. The cover was so thick out toward the end of a blade that sunlight was unable to filter through, and the chlorophyll-containing cells of eelgrass died and turned black.

The dense stands of eelgrass established a habitat at the bottom of the basin's inlet favoring large populations of highly mobile scallops. Each scallop rested upon one of its two fluted shells, the other crowded with attached red seaweeds, young mussels, and encrusting worm tubes. The accumulation of these other lives did not hinder a scallop's activity, for as soon as it detected movement with its two rows of bright sapphire blue eyes peeping from the fringed and tentacled mantle, it took off in a wild skittering dash, propelled by jets from its rapidly flapping shells. Wherever this happened, clouds of soft silt rose in the water to drift slowly through the waving eelgrass. As a scallop settled to the bottom, it remained closed for a moment, then opened slightly, its long, prehensile foot emerging to probe about on the valves, in the water, and down to the sand. If all seemed safely in order, the shells gaped wider, rows of long pointed tentacles emerging before the brilliant blue eyes were again revealed.

Scallops also took alarm at the frequent presence of starfish in the basin, but not by seeing their slow movement. They were keenly sensitive to chemicals secreted by certain cells toward the upper tip of each starfish arm. No matter how diffused these secretions might be, they were sufficient to send scallops into a frenzy of flapping, perhaps even breaking through the surface above before side-slipping down to rest again.

The presence of eelgrass beds was related to the frontal advance of a salt marsh, for sediment was trapped by the reduced tidal flow through the forest of ribbonlike blades. After years of accumulation, mud flats would finally be exposed at low tide, providing suitable ground for the germination of cordgrass seeds. The wide tidal basin, protected by its bar, had in fact decreased in depth over the years and was on its way to extinction as a completely aquatic habitat. In a century or two, it would be filled in and exist simply as an extension of the ever-growing salt marshes.

The sediment-laden water entering the bay encouraged a surge of growth activity, especially among microscopic plants and filter-feeding animals. Just inside the inlet to the basin, the water was darkened by masses of tiny, black tubes skittering about in a highly agitated fashion. Each tube contained a little cylindrical shrimp that extended two pairs of antennae to serve as organs of propulsion. A tube jerked along in one direction, then suddenly reversed its course going just as rapidly, for the shrimp could double back upon itself inside the tube in a fraction of a second. Despite being buffeted by tides, the huge numbers of tube shrimp had remained in this region for days, feeding on suspended organic remains.

Outside the inlet, the bay shore was extended by a brownish reef, exposed at low tide but now being covered as the water flooded in. Although wind-driven surf would crash over it occasionally, the reef, built of little more than sand grains, solidly held its position. It had been constructed by hundreds of thousands of honeycomb worms that picked sand grains of the proper size out of the roiling water and cemented them together to form the congested tubes in which they dwelt. Any one tube was fragile enough to be crushed or torn apart by predators or natural forces, but the combined effect was to produce a sturdy shoreline reef alternately exposed and covered by tidal bay waters. The worms inhabiting the reef had an elaborate array of appendages surrounding their heads, consisting of tentacles for capturing food, gills, and special tentacles and lappets for their masonry work. The broad head itself was surrounded by stiff bristles, the whole structure being used to plug the burrow's aperture when the tide went out, preventing moisture loss. As a further advantage, the spiked bristles discouraged predators from forcing their way in to get at the soft-bodied worm.

Since the reef had been established, an array of different animals found it a suitable and protected place in which to live, so the entire structure was a refuge for other, freer worms, small crabs, tube-dwelling shrimps, and a host of more minute plants and animals. The healthiest part of the reef was that which faced outward into the bay. Many of the tubes to the rear were no longer occupied by their builders and accumulated sediment at a rapid rate. Like the eelgrass community, the honeycombed reef was a potential site for salt-marsh development.

The flow of the river, augmented by the counterclockwise rotation of tides within the bay, was now beginning to be deflected from the western shore by a gentle curve that gradually increased in sharpness, finally terminating in a pointed cape that not only grew out from the mouth, but pointed almost directly north into the bay. In past centuries, there had been other capes in the same general region, but by their nature they were self-destructive. As they accumulated sand from the outer seashore, and picked up sediment from the bay, they grew long and ever more curved, until finally they doubled back upon themselves, creating a lagoon which soon filled in as marsh. The inner side of the present cape was frayed out in long feathery tendrils of sand that reached back into the bay, forerunners of the bar that would join with the shoreline a mile or two away. When this happened and the enclosed lagoon had disappeared, conditions would be right again for the beginning of a new cape.

Because the cape was composed of sand still in a highly active state, carried by bay and coastal currents and by prevailing winds, still another distinctive habitat developed for both land and marine animals. High on the crest of the cape, where dunes grew dark with organic matter and were stabilized by poverty grass, pitch pine, poison ivy shrubs, and fox grape, conditions were well on their way to becoming those of coastal lands. The soil was poor, but there were few signs of a maritime influence; only some of the woody plants showed the effects of sand-blasting and erosion by salt spray.

Lower down the dunes were not so stable, although pioneering marram grass stitched across open spaces where beach pea and a dune rose also succeeded in growing. The wide shimmering expanse of marram grass supported a few animals, nearly all of them predatory. Only one insect, a sand-colored maritime locust, ate the grass in any quantity, flying across the sand from one clump to another clacking its wide, conspicuously marked wings. A few meadow mice, making nests beneath old pieces of driftwood cast high during winter storms, also fed on the grass, cutting short lengths and packing them into their tunnels and chambers. Many more animals were carnivorous, including predatory beetles, digger wasps, velvet ants, and white wolf spiders that hunted for active or weakened prey. Hog-nosed snakes emerged from dense thickets as the heat of midday subsided to hunt for living food, perhaps in the form of young birds. The grassy dunes still held the nests of terns and black skimmers whose young had hatched, but were yet unable to fly and had to take refuge huddled against clusters of marram grass. As afternoon shadows reached the dunes, lightly mottled toads erupted from the sand, wiped grains from their eyes and nostrils, and hopped stolidly in search of insects.

Where the last barrier dune fell away to the beach, another predator was active—a white ghost crab whose deep burrows pitted the upper beach region. One by one they rose to the entrances of their burrows, shoveling sand that had accumulated during the day with their broad claws and doing a bit of repair work. Soon every burrow was surrounded by radial tracks of their pointed feet as they ran out to search for animal remains and bits of sea-weed. When alarmed, they dashed back to their burrows with extraordinary rapidity, ducked inside, then cautiously emerged, stalked eyes held high. A ghost crab was one of the very few animals that had made a direct and successful invasion of the land from the sea, not depending upon the migratory route of an estuary. In fact, an adult ghost crab could not swim, but returned to the waterline only to release its eggs which followed a typical larval existence before developing into creatures of the shore.

When the young had finally metamorphosed into tiny crabs, they left the sea and dug small burrows into the wet sand of the lower beach. As they grew larger with each molt, they moved higher, until the largest ghost crabs lived a hundred yards away from the water. A fully developed burrow of an adult was not a complex excavation, although it might descend four feet or more. Usually it consisted only of an oblique tunnel with one side entrance which was used infrequently for escape should the main entrance be blocked.

Wherever ghost crabs lived along the bay and ocean shore, they mimicked the background color of the sand. This was made possible by pigment cells beneath their hard outer skeleton that were affected by hormones circulated in the blood, the extent of these secretions being controlled quite involuntarily by what the crab saw around it.

One exceptionally large ghost crab paused in the mouth of its burrow, folded down its larger claw against the base of this appendage, and began sawing it back and forth. The base had a smooth elevated ridge, while the claw possessed a long, yellowish, toothed comblike ridge across its broad palm. As the toothed part of the claw was rubbed over the other, a distinct vibrating sound was audible, an efficient means of communication between animals with no vocal cords.

In the wrack zone, high spring tides had cast up rafts of driftwood, plant fragments, skate and whelk egg cases, and old shells. Crustacean beach fleas foraged here for worms and small snails, their white bodies almost invisible in the dim light except for bulging gray-blue eyes. The wrack supported many animals specialized to this temporary place of existence: mites, minute springtails, yellow beetles, long-jawed tiger beetles, ants, flies, and frequently their larval stages as well. Elongated, black earwigs used fragments of eelgrass to support the walls of their burrows excavated beneath rotting piles of rockweed.

The sand darkened below the wrack line, not only from moisture but from the organic sediment washed into it by every tide. Wherever small bay waves broke, the beach came alive with glistening, scurrying creatures that erupted out of the sand and just as quickly burrowed beneath again. Droves of brilliant bean clams appeared in the swash of a wave, each pair of shells colored differently from every other, while their long muscular feet probed into the sand and rapidly carried them from sight again. Mole crabs lived directly in the breaking waves, where they settled facing

the shore. As a wave raced up the beach and then began to
retreat, each mole crab unfurled a pair of long, feathery
antennae that filtered bits of food from the receding water.
When the tide level changed, mole crabs emerged and
dashed along with the surge, then with quick sculling
movements and a large, trowel-like abdomen, they backed
down into the sand again, keeping only their eyes and a
smaller pair of double antennae above the surface to serve
as breathing tubes. The long feeding appendages were kept
tucked away until the backwash carried food past the band
of mole crabs occupying this one level of the beach.

From the direction of the marsh, a shrill four-syllable
whistle preceded a willet flying low toward the cape shore,
its black and white wings flashing sharply in the afternoon
sunlight. It landed on the lower beach, its contrasting colors now hidden by a mottled sandy plumage, and stood
spasmodically bobbing its head up and down. After a
pause, the willet darted to the water's edge, where it ran
easily back and forth following the waves, glancing about
and taking in every sign of possible food. It was chasing a
retreating wave just as a mole crab backed into the sand;
with a quick probe of its long bill, the bird yanked out the
crab and ran back up the beach before putting it down,
jabbed it a few times, and ate it.

Just offshore, a rising, dipping cloud of sandpipers momentarily glinted white, then became dark and indistinct.
They flew as one, the whole flock passing with great speed
along the sand bars. Abruptly the birds wheeled toward
the beach and alighted on the wet sand facing into the wind
and spread over a distance of several dozen yards. Immediately they burst into action. After a long run, one would
stop as far down the beach as possible and probe quickly
for tiny crustaceans. As the next wave was building almost
on top of it, the sandpiper would turn and with twinkling
feet keep an inch or two ahead of the sheet of water sliding
up the beach. By the time the wave's momentum had sub-

sided, the sandpiper had already turned to follow it downhill again.

Not all life in the lower beach was so obvious. Between wet sand grains another world existed in which microscopic roundworms, copepod crustaceans, bacteria, and single-celled protozoans lived with near-impunity. Their existence was a quiet one until the violence of a sudden storm might wash a whole section of beach away, but their numbers were such that a newly exposed beach would soon be repopulated.

Under the low-tide mark, a trough ran parallel to the beach; it was inhabited by swarming schools of silversides seeking plankton trapped inshore by the waves, and by needlefish that had come close to shore to feed on these lesser fishes. Sand shrimp and spotted lady crabs burrowed into the bottom, above which smoking tendrils of sand were carried aloft by circular wave action a foot or two overhead.

In the distance, as twilight drew near, a pair of great dagger-winged black skimmers appeared flying rapidly along the shore just over the trough. Their white underbellies and bright red bills grew conspicuous as they approached. Each bird held the lower, longer half of its beak in the water, leaving behind a V-shaped, furrowed wake. As they crossed the school of silversides, the skimmers snapped up two or three fish in rapid succession and continued their rapid flight up the bay shore, one of them crying out raucously before disappearing around a slight promontory.

The silversides, thoroughly alarmed, fled away from the beach. As they crossed shallow water over a sand bar, a half-dozen circling terns braked in mid-flight, hovered, and plunged into the water, scattering the school. After each dive a tern reappeared above the surface with a gleaming three-inch fish dangling from its bill.

The bar, running offshore toward the cape, supported its

own distinctive populations, mostly composed of burrowing animals. Brilliant, metallic-hued plume worms stretched
out from their chimneylike tubes that were fortified and
decorated with shell fragments, pebbles, seaweed, and bits
of hydroid stems. The worms leaned down to the sand
bottom some inches away and swept about in a half-circle,
collecting food that had come to rest nearby. Now that the
tide was flooding, great lugworms, a foot beneath the bottom, swallowed sediment-laden sand by means of a ballooning proboscis thrust out and inflated with body fluids.
After a dozen or so swallows, each worm would rest for a
while before beginning to feed again. As earthworms of the
sea, the lugworms periodically cast upon the bottom ropy
piles of partially digested wastes that added to the organic
matter of the bay floor. Trumpet worms, occupying conical cases constructed of carefully selected sandgrains indi-
cating the finest masonry work, dug head downward in the
muddy sand, using golden bristles surrounding their flat
heads to excavate or to plug their cases in times of danger.
A large white parchment worm ensconced deep in a U-
shaped burrow was so specialized in form it was little more
than a complex, segmented series of waving valves and flaps
that produced a vigorous, one-way flow of water. In its
dark tube, the worm glowed brightly, luminous in its entire
length, although the light would never be seen from the
surface. Every now and then the parchment worm was
nudged a little by a small colorless pea crab also living in
the tube, but the worm did not respond. The crab was in
no way harmful to its host, but found complete security
and a continuous supply of food in the well-flushed tube.

The way of life of the parchment worm was not unique
in the sand bar, for an entirely different kind of creature, a
white burrowing ghost shrimp, entertained its share of
free-loaders too—pea crabs, scale worms, an occasional
slender fish, and other smaller creatures. The elongated
ghost shrimp was more industrious than any of the worms,

constantly keeping excavations underway with its first three pairs of legs and heaping mounds of sand outside the burrow. It too was a filter feeder, devouring plankton and organic particles strained from the continual flow of water by means of fine hairs on its claws and other appendages.

Oval holes in the rippled surface of the sand bar led to the siphons of razor clams beneath. The long slender molluscs filtered food from tidal currents too, but were not restricted to one spot and in fact could be very active. Should one be uncovered by strong currents, or uprooted by some bottom feeder such as a skate, it could dig down into the sand with its daggerlike muscular foot more rapidly than most other animals could follow.

The sand bar, also the home of hermit crabs, flounders, and pipefish, gradually fell away to the bay floor which was composed of a pavement of shells, sticky clay, or fine sediment, depending upon the sorting action of currents. In regions where the bay was carpeted with sediment that was almost half fine silt, the preponderance of life burrowed into the bottom; in other places, digging was difficult and the animals present were mostly those that lived on the bay floor without penetrating it. Taken as a whole, the bulk of bottom life in the lower bay was of the burrowing kind, mostly clams and worms that were safe from strong tidal currents and the inevitable rain of sediment. Along the western wall of the bay, where the southward-flowing river current streamed past the cape, much of the sediment was comparatively new, having been carried down from mountains far inland. Elsewhere the mouth of the bay was either washed bare or carpeted with sediment carried by tides from the continental shelf outside the capes. This too had a river origin in times past, but had been a part of the shallow sea bottom for a long time.

For the first time, the river was almost blue in the region of the lower bay. Brooks had been crystal or tea-colored; streams and rivers mostly yellow-brown; the ox-bow ponds had been green tinged with yellow; and the estuary and upper bay a distinct brown. Except for the color of brooks, which was a stain from leaves and pine needles, the other colors had been the result of light interrupted by suspended particles in the water. Now that most of the sediment was settling to the bottom of the bay, the water was freer of these living or inanimate particles which slowed sunlight and turned it toward the red end of the spectrum. The bay was still a far cry from the dark pervading blue of the open ocean which was, in effect, a marine desert, but it was blue nevertheless.

At the mouth of the bay, nearly all vestiges of a fresh-water influence had disappeared and the creatures present were almost exclusively of marine origin. Plankton organisms were more varied and less populous than they had been upbay. The stinging sea nettle jellyfish common in the bay was less common, its role having been taken over by larger, dark-red sun jellies and disk-shaped moon jellies. Where large washed boulders littered the bottom, tall red and brown sea anemones stretched their tentacles into the tidal currents; bright orange, treelike whip corals swayed three or four feet off the bottom; and round clusters of pink northern coral nestled among the rocks at the bottom. A limited population of northern rock lobsters lurked under these same rocks, their heavy, asymmetrical claws held in readiness for the destruction of a weakened or dead animal. This was the southernmost extent of their range, but they were active and well-established although far smaller than members of their race several hundred miles north of the bay.

The ancient river channel had not been filled in, but was scoured to a depth of one hundred and fifty feet by the powerful thrust of water leaving the bay. With every flooding or ebbing tide, the deep channel was a scene of chaotic activity as clusters of seaweed and plant fibers rolled bouncing along like desert tumbleweed. This nearly lightless region was heavily populated by all kinds of intruders from the ocean, for here they found abundant nourishment flushed from the bay shores and equable temperatures only slightly affected by seasonal changes. The deep channel was an excellent feeding ground for predatory fishes able to capture other fishes and the multitudes of molluscs, crabs, and worms. The motionless figure of a grotesque angler fish, its lure dangling over a toothy mouth almost as wide as the fish was long, seemed an indication that the gateway to the sea had been reached at last.

The land dropped away, but the river flowed on in still greater volume, hidden in its complex patterns of joining the sea and no longer confined by the continent it left behind.

IX

THE RIVER
SET FREE

The river is within us, the sea is all about us;
The sea is the land's edge also, the granite
Into which it reaches, the beaches where it tosses
Its hint of earlier and other creation. . . .
                                          T. S. ELIOT

THE low capes faced each other across the bay under a
gray sky, constricting the river's mouth. In large measure
they directed the strong tidal currents that flooded in,
fanlike, along the northeastern side of the bay. The tides
later rushed out, augmented by river flow, to keep the
channel deep and free along the southwestern shore. Here
was the end to the river's confinement, yet the channel did
not become indistinct beyond the capes; it retained its
physical identity long after emerging upon the wide continental shelf that stretched out beneath the surface of the
blue-green ocean.

A great cloud of water from the bay billowed out into
the clearer waters of the shelf, mushrooming beyond the
capes in ever-changing patterns according to the vagaries
of tide and current. At the surface, its slow progress appeared confused and disorderly. Nevertheless, its course
was determined by a certain regularity of the coastal currents. Directly off the southern cape, a small clockwise

rotation of water brushed against the shore, extending the cape northward year after year, transporting sand placed in suspension by waves breaking along beaches south of the bay. The entire coast facing the ocean had been receding for many centuries because of this continued loss of sand. Tree trunks, now half-buried a quarter of a mile offshore, had fallen into the sea three hundred years ago when they had been undercut by waves breaking upon the shore of that time.

The internal cylindrical rolling motion of waves was distorted when they entered shallow water and approached the beach, their forward momentum causing rising crests to topple forward as foaming breakers. Because they washed the beach obliquely, the shoreline was molded into sculptured cusps as wave and backwash transported sand toward the cape. At intervals surplus water driven upon the beach accumulated from the pressure of the longshore flow and rushed seaward in a powerful rip current. By the cape, where the river crossed tidal currents, the whole sea surface boiled in white turbulence as rips swirled with the friction of opposing forces.

Ten miles off the bay, a large circular current also revolved in clockwise fashion, separated from the one along the shore by another turning counterclockwise. Up and down the coast, a series of these rotating currents were acting like great fluid gears coming in contact with one another. Through friction, one turned the next in the opposite direction. Plankton and floating debris were transferred from one huge eddy to another, often being carried many miles along the coast. All these inshore rotating currents derived their main impetus from a belt of water drifting southwest over the shelf, itself driven by an elongated circulating current just outside the shelf. This was the last gear in a system that transferred energy of motion from the single huge ocean current flowing to the northeast. It was this enormous oceanic stream, set in motion by tropical heat and directed by the rotation of the earth and its restricting continents, that had carried larval eels northward and had created conditions necessary for the river's origin months before in the distant southern sea. The whole circulation pattern was one of great oceanic conveyor belts and gears, of energy transferred from one moving mass of water to another. The still-turbid river flow seemed to disappear amid a welter of streaming and rotating currents, some of its transported particles drifting downward to add to the shelf and others carried in almost permanent suspension throughout the ocean. It was not at the surface that the river's continuing influence was most strongly felt, but along the sloping bottom of the continental shelf in what remained of the channel.

The descent of the shelf was very gentle, in most places only a dozen feet a mile, but its vast submerged plain extending one hundred miles from shore was convoluted by elevations and depressions, sandbanks and mud flats, and great jumbled heaps of boulders—a topography as varied as any on land.

A narrow shelf had existed in preglacial times, but it was during the four separate ice ages when glaciers advanced and retreated, eroding and later depositing sediments, that the shelf had developed into its present form. At the onset of each glacial epoch, ice had crept down the continent and over the shelf, at first floating out as a wide, thin sheet, then freezing all the way to the bottom where it scraped and scoured the shelf deposits exactly as it did on land. The preglacial shelf had been composed mostly of fine river sediments dropped directly into the sea, but after each ice invasion it was littered with cobbles and boulders, gravel, silt, and wide expanses of sand with only an indistinct pattern to their distribution. Silt and sand from swollen glacial river deltas occupied most of the central and inshore areas of the shelf, while stones and larger rocks fringed the shelf margin where they had been left by melting tongues of ice, after having been pushed along for hundreds of miles by the advancing front. When the glaciers had finally gone, most river-carried sediment fell into the newly created bay, leaving the shelf almost intact as evidence of its past history.

During the few thousand years when the sea had been at its lowest, at least three hundred feet below the present level, shoreline waves had cut terraces into the narrow belt of soft sediments left exposed. When the glaciers began their retreat, they did so rapidly for a period of ten thousand years, water from their melting causing the sea to rise with equal rapidity. Occasionally the rise would slow enough to allow other major terraces to be formed before the sea commenced its ascent up the slope again. This sporadic opportunity for coastal waves to cut deeply into the loose sediment resulted in a series of steps, not too well defined any longer because of subsequent depositing of loose sand.

One of the more obvious aspects of the shelf was a series of submerged elongated hills running parallel to the shoreline. They were underwater counterparts of the present barrier islands lying just off the coast, which sheltered quiet lagoons that were slowly being invaded by salt marsh. Because the rise in sea level had not been continuous but had halted periodically, there had been sufficient time for these earlier barrier islands to develop fully before being submerged when the sea level rose again.

All these features were superimposed upon the submerged hills and valleys that had been created directly by glaciers. The topography of the shelf was not unlike that of the exposed coastal plain and was merely an underwater continuation of the same characteristics which distinguished the alluvial plain from the rocky piedmont far inland.

From age to age, the depositing of sediment upon the shelf had kept pace with the slow sinking of the entire coast

under its increasing weight. Deep within the shelf, evidence
of this subsidence existed: the older and deeper the layer of
sediment, the more it tilted toward the sea bottom. At the
very base of the outer shelf, nearly twelve thousand feet
beneath the present bottom, the original slope ran directly
from the ancient shoreline at the edge of the piedmont to
the abyss of the sea floor. The massive layers of sediment
extending above this base of solid rock were evidence of
the contribution of the river, its predecessors, and its neighbors. The sculpturing by glaciers and coastal currents that
resulted in the contours of the present shelf had been done
comparatively recently in the earth's history.

Quite apart from its original deposits and a continuing
gentle rain of current-borne particles, the river still played
a major role on the shelf. Its deep channel emerging from
the mouth of the bay began to lose its sharply defined form
and followed a moderate, shallow depression that marked
the river valley of earlier times when the sea level had been
much lower. It proceeded straight across the shelf, encoun-
tering many basins on its way to the edge where the steeper
slope that plunged into the abyss was eroded into a precipi-
tous canyon. Before the submerged river valley reached
this canyon, however, it splayed out in a wide deltalike fan
of alluvial deposits and rills, with one or two major channels continuing down into the deep cleft.

Although its diffused flow persisted, the river's character
had changed. It exhibited only the faintest traces of fresh
water; drop for drop, it was chemically almost indistinguishable from adjacent coastal sea water. The river continued to flow across the shelf in its ancient bed only
because it was now more dense than ordinary sea water
from the heavy load of sediment it still transported. Its
increased weight caused it to slip downhill, seeking the path
of least resistance, in much the same fashion it had done in
the hills far away.

Deposits making up the continental shelf differed from

those lining the river estuary and carpeting the bay, although most of them had in fact originated on land. Sand still was carried down as it had been for thousands of years, its quartz and feldspar crystals and thin mica flakes glinting in the subdued light. With time, there had been additions of other kinds of sand from quite different sources. Some particles were precipitated directly but very slowly from chemical compounds in the water. A more abundant kind of sand was composed of the near-microscopic shells of one-celled foraminiferans and of tiny crumbled fragments of larger shells, corals, limy worm tubes, and coralline plants. Interspersed with these were broken sea urchin spines and needlelike sponge spicules. All three kinds of sand—mineral, chemical, and organic—were thoroughly mixed together and held by cohesive muds and clays. The muds that bound together much of the looser sediment were very dark, with faint tinges of red, blue, or green, depending upon the way in which they combined with oxygen and silicon. Often they were heavily viscous and difficult for burrowing animals to penetrate.

There was little regularity to the bottom, although washed sand usually covered the wide, flat areas. Mud commonly collected in depressions, while rocky gravel lay exposed on the crests of the low underwater hills. Generally these patterns remained quite stable, but a major shift in currents could disrupt the transitory nature of shelf topography at any time.

Life in loose, sand-and-mud bottoms had to contend with conditions substantially different from those in open water. Water penetrating the spaces between sand grains had an altered chemistry due to a limited circulation, prolonged contact with certain minerals, and an accumulation of wastes from living organisms. Water within the substrate was deficient in oxygen and increasingly alkaline; it contained more nitrogen wastes than would normally be present in sea water. Populations of those organisms with little or no contact with the marine world above were limited mostly to bacteria, roundworms, and small burrowing segmented worms. Closer to the interface, tiny elongated crustaceans of a sort virtually unchanged for millions of years tunneled through the loose sediment, feeding on deposits of organic matter. These round-headed little creatures had no close relatives in the world today, but resembled an ancestral stock from which all crustaceans sprang.

Sandy areas of the shelf supported a host of burrowing animals just below the surface. Very large, heavy-bodied clams lived in close proximity to one another, buried except for their siphons, which drew in steady streams of water to be filtered for plankton. Sand inevitably was sucked in, but the clams trapped the grains in strands of mucus that flowed out between the shells, removing the abrasive particles from easily damaged soft tissues. Beds of these clams often covered several acres, crowding out most other burrowing forms that might compete with their filtering activities. Wherever the clams lived in quantity, they were the dominant forms of life; where the bottom was not so suitable, their populations diminished and other molluscs, worms, and burrowing crustaceans appeared in greater numbers.

A contentious burrowing mantis shrimp lived in territorial isolation in loose sand, not tolerating others of its kind nearby. The eight-inch shrimp was grotesque, with a heavy muscular abdomen and a pair of folded jackknife claws held in readiness just beneath the sand. When a smaller creature happened by, its claws shot out, unfolding with lightning rapidity to seize the victim. Escape was nearly impossible, for the inner edges of the hinged claw blades were serrated in long sharp spines. The presence of approaching prey was detected partly by an acute sense of touch located in both the animal's antennae and in sensitive bristles on the body, and partly by elongated, mallet-shaped eyes held above the sand on jointed, ringed stalks totally unlike those of any other crustacean.

Without warning, the mantis shrimp erupted from the sand, shedding particles from its spiny body as it swam powerfully to a new location. It moved through the water smoothly, propelled by paddlelike appendages beneath its large abdomen, the rows of flaps moving in sequential fashion down the long body. During this exposed transfer to a new site, it was reasonably safe, for nothing but a very large predator could cope with the long, exceedingly sharp spines that emerged from almost every part of the mantis shrimp's jointed external skeleton. It descended slowly to a bare patch of sand, settling beneath the surface in a flurry of fanning and burrowing movements of its many appendages, and waited silently.

Close to the mantis shrimp, but quite safe because of its size, anatomy, and way of life, lay a sea mouse, one of the strangest of all segmented worms. It was several inches long, oval and thick of body, its upper side densely covered by a feltlike mat of brilliantly iridescent bristles. The color was lost to animals swimming by, for the worm burrowed head downward into the muddy sand, leaving only its blunt rear end exposed at the surface. The worm drew in water underneath its body by rhythmic contractions, then passed it back along the top side under rows of overlapping scales and the coat of fibers and bristles that covered them. Oxygen was extracted from sea water in this way as the sea mouse crept through the loose bottom searching for prey. When it came across a smaller worm, it suddenly everted a muscular pharynx armed with sharp teeth. Not only was this pharynx an efficient trapping device, but it served as a gizzard to grind up whatever food was captured.

A neighboring creature, a long and slender lancelet, backed slanting into the sand, its head protruding from the surface. It sucked in water through a funnel-like mouth that was surrounded by short tentacles. The flow passed down a gullet, into a chamber, and through a series of slits on either side of the body where microscopic plant food was trapped in bands of mucus. The filtered water then left through a vent underneath the lancelet's body. This feature, together with the placement of a main nerve cord along its back and a stiffening rod just beneath the nerve, linked the lancelet to fishes and all the higher creatures of sea and land, although it was far simpler than they. It too had evolved from a common ancestor all shared, but it clearly retained the characteristics of those ancient sea creatures that had formed the basic pattern upon which all vertebrates had built over many millions of years.

The sand harbored relicts other than the little elongated crustacean and the lancelet. A flattened lamp shell lived in a burrow, anchored to the bottom by a very long muscular stalk, this particular form almost unchanged in its five hundred million year history. The lamp shell—not in any way related to molluscs—was another example of the haven the sea offered to a few animals surviving from the past. There were only two hundred different kinds of lamp shells in the oceans, yet three hundred million years ago there had been over thirty thousand distinct types, most of them congregated in huge populations in shallow coastal waters.

Nearly every basic type of organism that had existed in the earliest days of life on earth still had a few representatives living in the present sea, which served as a repository of still-functioning evolutionary experiments. Only a handful of entire plant and animal groups had vanished into the oblivion of extinction, incomplete fossils the single remaining evidence of their previous existence. For a very few groups there were not even fossils—just traces of faint, wandering trails left in mud hardened to rock.

Over millions of years, each organism, plant or animal, ancient or modern, had developed specialties that allowed it to exist with little direct competition from other species. In the event a contest did arise between two kinds of organism, both striving toward precisely the same means of existence, only one of the competitors would survive, the less successful either disappearing permanently or its diminished populations finding a new and uncontested specialty. The exploitation of a new way of life would depend upon the successful emergence of inherited traits present in only a relative few among the endangered population. Patterns of interaction in the sea were complex and ancient, established by being a plant, therefore a primary producer of life substance, or by being grazer, scavenger, or predator. The coastal sea in all its environmental latitude not only preserved numerous ancient ways of life, little changed from the time of their first appearance, but was a showcase for some of the most advanced animal forms, such as the dolphins that played near the rip current beyond the mouth of the bay as they headed for deeper water.

The bottlenose dolphins, now rolling slowly through the long offshore swells, had distant links to the rivers of the world, for it was up such waterways that the fishlike ancestors of all terrestrial animals with backbones had migrated several hundred million years ago. After long ages as air-breathing fresh-water fish, amphibians, reptiles, and finally mammals, some warm-blooded creatures began leaving the land for a more aquatic life, a few eventually departing from fresh water and descending rivers and estuaries to take up a marine existence. By then they had become so thoroughly adapted to life in water that many of their terrestrial traits had vanished. Throughout the river and sea an entire range of mammalian specializations from land to water was evident, each independent of the others. Mink and otter were equally at home on land or water; muskrat and beaver were more aquatic than not. Elsewhere in the world, sea lions could still gallop across rocky beaches on strong fins, but the rotund earless seals sometimes found off the river's mouth could only roll up on low-lying rocks and inch along by strenuous effort, their hind legs unable to be turned forward. About thirty million years earlier, both kinds of seals had emerged from a strong-legged, tailed ancestor with aquatic habits, very possibly an otterlike animal.

A much older migration to the sea, at least twice as old, provided the basis for dolphins and whales of the present, creatures evolved so completely into a marine existence that they could not survive being stranded on a sand bar by a receding tide. Buoyed by water their whole lives, their internal skeletons had lost much of a former capacity for support on land and were unable to keep lungs from collapsing in all but the smallest porpoises and dolphins. Ribs had diminished, hind legs were gone, and the long flexible backbone served mostly as a fulcrum for the powerful trunk muscles which drove the animal forward with vertical thrusts of widespread flukes at the tip of the tail. Nostrils now were situated at the top of a dolphin's or whale's head and were provided with valves, below which the respiratory tube completely by-passed the throat, making it possible to eat and breathe at the same time. These were anatomical improvements that had developed very slowly. The earliest whales, preserved fossilized in rock, had nostrils placed far down their elongated snouts, close to the position of land mammals. Instead of hair, which had disappeared, thick blubber in modern whales provided protection against the cold, while a smooth rippling outer skin in the dolphins created a thin layer of turbulence that so reduced friction with the water as to place these animals among the swiftest in the sea.

The dolphins were seeking food, relying upon an extremely sensitive echo-ranging system. They emitted bursts of high-frequency sound which radiated into the surrounding sea. Whenever fish were encompassed by this shell of sound, their reflections sped back toward the dolphins, causing the mammals to alter course directly toward their prey. The dolphins were seldom confused using this echo-location sense, for it was so exquisitely detailed that one could instantly determine the size of the fish or its school and the direction being taken. The large, convoluted brain of a dolphin with its packed layers of nerve cells that received and associated myriad impulses had no equal among the four-legged animals of land.

The dolphins continued their slow roll in and out of the waves, several of them trailing long parasitic barnacles from their curved back fins. They were communicating back and forth through a different and lower frequency of squeaks which rose in an excited chatter as they located a school of herring. The dolphins spread out, each animal thrusting powerfully with its flukes, attaining a speed few creatures in the ocean could exceed. They flashed into the alarmed herrings, their peglike teeth unerringly gripping one fish after another as the echo-location came into clear focus.

Far below, the mantis shrimp's eyes were not the only ones protruding from the sand. Across the wide expanse of shelf other eyes watched, some flecked with gold, some shielded by fingerlike lappets, and none suggesting the size or kind of animal they served. The goggling, close-set eyes of summer flounders were common, as they had been in the bay. A different pair of eyes looked directly up into the blue-green water above, with only a little patch of pale skin revealed between them. The patch was important, for it was capable of discharging an electric shock from the bulky, club-shaped stargazer, to which it belonged. The fish's electric organ was primarily a defensive apparatus and served warning to leave the delicate eyes alone. The stargazer could swim well enough, but was so thoroughly adapted to life on the bottom that its mouth and gill openings, as well as its eyes, were arranged to be at the surface. The mouth, a wide slit, armed with sharp teeth, was a formidable trap for any creature that ventured too close. When a small fish came by, picking at bits of organic matter on the bottom, it suddenly disappeared from the open water of the sea, sucked in by a violent inhalation of the fish's capacious mouth.

Behind another pair of eyes, each shaded by a scalloped lobe hanging down over the curve of its glassy cornea, sand fluttered in little swirls as round gill openings appeared and closed rhythmically in what seemed to be a smooth bottom. A faint outline where the fish had buried itself revealed the presence of a clearnose skate that had recently arrived, its wide diamond-shaped body, mostly composed of fins, rippling smoothly. When it had settled to the bottom, the same rippling motion, now more vigorous, had cast sand up and over the body; with an abrupt wriggle of its long, spiny, finned tail, the skate had disappeared into the bottom. Unlike the stargazer, its mouth was on the underside of its body as befitted a specialized bottom feeder. Whenever a worm shoved its way through the sand, or the slow movements of a clam were felt, the skate's protrusible jaws dropped down and the victim was drawn into a mouth paved with broad and rounded crushing teeth. As it lay there, a young blue crab swam by, sculling itself over the bottom and quite unaware of the two golden eyes watching its progress. When it was within reach, the skate burst from its hiding place and with a furious, convulsive flapping, leaped over the crab to hold it imprisoned against the sand, while its jaws reached out to seize and crush the crustacean.

Other broad, cartilaginous relatives of the skate pursued more active lives. Stingrays soared from one spot to the next, feeding on the multitudes of sand shrimp hidden in the top half-inch of sand and on bloodworms as they pushed through the same layer. A large school of cownose rays, each over three feet wide, swam overhead in slow and graceful formation. At first they were so near the surface the pointed tips of their winglike fins broke through into the air. When they passed over one of the extensive sandy areas, all veered downward in a wide spiral to the bottom sixty feet below where they used their curious, blunt heads as shovels to extract moon snails and clams, or to break mussels free of their attachment to piles of glacial rocks. Hermit crabs scuttling across the bottom were captured repeatedly, their shell houses offering no protection against the powerful jaws of the rays. Across the shelf, other groups of cownose rays were gathering; before long all would band together in one enormous school stretching for miles as they made their slow way south with the approach of fall.

Skates and rays had evolved from elongated sharklike ancestors more than a hundred and sixty million years before, gradually becoming flattened with broad enlarged fins until a diamond shape resulted. Rays swam like birds, their great fins gracefully beating up and down, but when near the bottom and needing to move only a little, such wide sweeps were not efficient and stirred up clouds of sediment, so they adopted the technique of skates. With precise muscular control a ray could ripple the outer margins of its fins in sinuous, rearward waves, until the large bulk of the animal was lifted off the bottom and propelled gently forward.

Still other cartilaginous fishes had not achieved the flattened body of a skate or ray, neither were they as slender as a shark. One such fish lying on the bottom, a torpedo, had a rounded broad body with a short, finned tail. Like the stargazer, it had the capacity to generate electricity in highly specialized cells developed from nerve and muscle, but its shocks were considerably more powerful than those of the smaller bony fish. Electrical properties are inherent in all life; every living cell possesses positive and negative charges that are essential in maintaining its chemical equilibrium, especially in the membranes surrounding a cell. During the evolution of many electric fishes living in fresh

or salt water, a gradual accentuation of these properties of
nerve and muscle had taken place. Embryonic cells that in
other fishes became muscle cells, in the torpedo had gone
into the construction of electric cells. Each cell was capable
of discharging only a minute amount of current, but the
cumulative in-series effect was enough to stun much larger
animals, especially as the shocks could be repeated over one
hundred times a second for a short period. That its electrical capacity was not merely a defensive measure was borne
out by the single large torpedo lying half-buried in the sand
near a cluster of rocks. Only minutes before, it had leaped
upon an adult flounder that had attempted to pass by,
stunned it, and eaten it whole. This was prey the ordinarily
sluggish torpedo would not be able to capture by any other
means.

The fish population living in waters over the shelf was
largely predatory: large fishes ate smaller ones and they fed
upon shrimps, molluscs, worms, and tiny creatures of the
plankton. Eventually, at the end of every chain, an animal
was nourished by the plants it ate, plants being the only
original producers of life material with its locked-in energy
derived from the sun. The first animal in an extended series
might be small enough to graze on microscopic diatoms; it
could be large enough to eat seaweeds; or it might depend
upon organic silt, mostly plant in origin, on the bottom or
filtered from the water.

Lines of predation were seldom clear-cut. Crevallé jacks
—deep-bodied silvery fishes—did not eat only the small
crustaceans among the rock piles that were their favorite
food, nor did black drums feed exclusively on mussels
attached to the bottom. Both ate a wide variety of other
creatures as they wandered over the shelf. The more vora-
cious predators swiftly caught any animals within their
ability to do so. Sleek young bluefish and northern barra-
cuda slashed into shimmering schools of sand lance, capturing the agile slender fishes before they could escape. Larger

bluefish hung around massive schools of menhaden, taking their pick without difficulty. At times a hammerhead shark would thrust through the packed menhaden, eating its fill, but an even greater slaughter occurred when a long-tailed thresher shark appeared, circled through the school in ever-decreasing circles, whipping its scythelike tail to concentrate the fishes for easier capture. It then turned and scooped up dozens of the crowded, stunned menhaden with little additional effort.

Over one of the rock-capped ridges where they had been feeding, a school of butterfish was interrupted by a lithe blue shark swimming straight up the slope from deeper water off the edge of the shelf. Under its belly it carried two remoras which detached briefly to capture a few of the smaller butterfish before hurrying back to their living means of transportation. A remora was a perfectly adequate swimmer, yet it depended much of the time upon sharks, large fishes, and sea turtles for its travels, attaching by means of a highly modified dorsal fin that extended forward over its head. The fin had become a flattened, louvered apparatus capable of exerting a powerful suction grip when muscles contracted to elevate its many transverse plates.

A little striped pilot fish swam busily in the shock wave just in front of the shark, occasionally darting out to secure bits of food. As the shark tore apart some of the butterfish, both pilot fish and remoras fed on fragments that escaped its jaws. When the blue shark had come up over the rise, its head had been sleek and streamlined; now that it was feeding, its whole jaw apparatus had dropped down and protruded forward. Slicing through a butterfish required little effort, but there were times when the shark attacked much larger prey. Then it would grip its victim securely with rows of pointed teeth and roll and twist its body, shaking in a frenzy. Other teeth, triangular with serrated knife edges, easily cut chunks of flesh from the inert victim which, though still alive, was prevented by resistance to the water from moving with the excited shark.

The blue shark and all of its cartilaginous kin were only distantly related to the bony fishes that made up so much of their food. They were not primitive leftovers from bony-fish evolution, but highly specialized animals sharing the same fresh-water ancestry as the other, more populous fishes. Bone had first developed in several groups of armored fishes living inland nearly four hundred million years ago. Less than fifty million years later, both bony fishes and sharks had embarked upon their separate ways, the cartilaginous skeleton of the latter being an embryonic characteristic retained into adulthood.

Life was more intense where rocks crowned the submarine ridges than on open sandy areas. If the exposed rock surfaces were high enough to reach into the more strongly illuminated levels of the sea, they supported thick stands of red and brown algae swaying only slightly in the subdued motion of waves that passed thirty feet overhead. Elongated pipefish and cornet fish drifted inconspicuously in the masses of algal filaments, while dark sea horses wrapped prehensile tails around the plant stalks.

A sea horse was not as efficient a swimmer as its close relative, the pipefish, but had to rely upon a heavily vegetated bottom to cling to. When one did swim, it was primarily by means of the rapid fanning of delicate, transparent fins on its erect back and at either side of its neck. In other ways the two fishes were similar. Each had mobile eyes which goggled about in search of food; each sucked in this food through a long pointed snout with small jaws at the very end; and each was tough, stiffly flexible in a coat of bony armor. The most unusual characteristic they shared was that of reproduction, for the females deposited their eggs in capacious brood pouches of the males where fertilization took place. There the embryos grew, nourished by blood vessels lining the pouch, until it was time for their birth, an event at present taking place in one male sea horse which was releasing over four hundred young in a series of vigorous contractions of his pouch. Many of the tiny sea horses clung to the male, giving him a fuzzy appearance, but most drifted away into the seaweeds where they might have a chance to survive.

The rocks were coated with a wide variety of animal life as well as plants. Whip corals, encrusting corals, and sea anemones studded many of the boulders, their tentacles reaching out to capture planktonic food or larger animals that passed by. Sponges of different colors, but mostly red and yellow, grew luxuriantly on the vertical sides of rocks and deep within crevices and holes. Colonies of reddish sea squirts—sea pork—formed lumpy masses and other sea squirts, looking like tangled strings of green beads, flourished among the bases of brown seaweeds. Their greenness was not due to chlorophyll-containing plant cells incorporated in their tissues as certain green shoreline creatures had possessed, but to the presence of the metal vanadium, an element so rare in sea water as to be almost unidentifiable. Somehow, during the filtering activity of the green bead tunicates, the scarce substance was retained in their blood as an oxygen-capturing agent, serving in this capacity as did the red iron pigment in the blood of worms and fishes and the blue copper pigment in some of the larger crustaceans.

The jumbled pile of glacial rocks provided safe haven for an almost endless variety of motile animals that crawled or swam across the bottom. Inflatable puffers and burrfish, not very strong swimmers at best, ducked in and out of the wider apertures. Lobsters and crabs picked their way through the labyrinth of crevices, exploring far into the darkness but avoiding one spot where an eight-foot conger eel lay hidden during the day. This formidable, gray relative of the common eel that had penetrated the bay and river emerged at night to feed upon the many fishes and crustaceans seeking shelter as their own activity decreased in the waning light. In the dim hours of twilight, the conger eel's large eyes assisted a keen sense of smell in locating food. Once its powerful jaws with their several rows of closely set teeth clamped around a victim, no escape was possible.

The rocky ridges increased in size and number as the edge of the shelf was approached. Here, where the water was deeper, many oceanic creatures ranged up and down the continental margins, at times striking out into open sea toward other land masses far away. It was here that a school of several hundred bonitos swam effortlessly through the clear water. The bonito, an example of almost perfect streamlining, was capable of sustaining high speeds over long distances. Unlike most other open-sea fishes, it lacked a gas-filled swim bladder within its body, so had no buoyancy and would sink were it not for its ability to swim constantly through its entire life. Because of streamlining, comparatively little energy went into propulsion and the fish glided forward in the smoothest and most efficient manner possible. There were no interruptions to the water flow over the major part of its body: its conical head, with tightly closed jaws and flat gill covers, offered little resistance. Even its eyes were set flush in the head, without the bulging corneas of most fishes. Its spiny fins retracted into grooves and a film of mucus covered the minute scales on its body, lessening friction to a minimum. Only small soft fins flared out toward the rear of its trunk to serve as stabilizers, while a series of tiny finlets interrupted the smooth flow of water to create a turbulence behind the tail fin that almost completely eliminated drag. The sole propulsive device was the crescent-shaped tail fin, set in motion by powerful, alternating contractions of banded muscles on either side of the axial backbone that swept the body back and forth in a wide arc. Only a few fishes, among them the closely related tuna and albacore, could match the efficiency, speed, and endurance of the bonito.

The course of the bonitos suddenly changed as they fled swiftly toward the distant shore, but not before one to the rear of the school was caught by a huge white marlin swimming even more rapidly in a tremendous surge of power, its short pointed sword cleanly cutting the water in front of the massive fish. Another bonito was caught and still another before the marlin swerved toward deeper water again. It rocketed into a school of round herring, catching at least a dozen before it began to slash about with its sword, cutting and stunning many more which it then fed upon more quietly as the remnants of the school of herring escaped. Later it pursued some of the most elusive of sea creatures, squids that shot along by jet propulsion, their tentacles trailing behind. But even they were no match for the marlin and many were caught and swallowed whole.

The squids had also been pursuing herring with considerable success. The conical body of each squid was in effect a muscular, inflatable chamber surrounding an inner body consisting of stomach, heart, and other organs. The pumping action responsible for its rapid progress began when water was taken in through a mantle collar around the neck, whereupon the squid would pull in its head, locking it in a pair of cartilaginous tongues and grooves against the collar of the conical mantle. The entire mantle would then contract, forcing a high-velocity stream through a tubular siphon beneath the head, the siphon's position determined by muscles. In effect, the squid had a powerful jet device that could be pointed in various directions. When feeding, it often proceeded with its eight arms and two tentacles foremost, but always reversed its direction for high speed escape maneuvers, the pointed, finned body piercing the water with little resistance.

The squid was totally unlike all other molluscs except its near relatives—cuttlefish, shelled nautiloids, octopus, and deep-sea forms; it could hardly be confused with snails or clams. It possessed huge, highly developed eyes of an acuity equal to any fish's, yet evolved independently. Its mouth bore a pair of sharp, black jaws that could shred a small fish or neatly remove a tentacle from another squid in conflict. Inside its rounded head was a large and well-formed brain, protected by an internal skull, and to sustain its high degree of activity the animal had three separate hearts to boost its circulation. The outer delicate skin of a squid bore hundreds of color sacs, each controlled by muscle cells which caused the sacs to expand and contract. Waves and patterns of red, yellow, and brown passed along the body, especially when the animal was aroused. As the marlin caught up with the school of squids and their evasive actions were unsuccessful, they emitted explosive bursts of sepia ink which clouded the water, but only a few escaped as a result.

Far beneath this action, many other squids were engaged in egg-laying. Fertilization had occurred several days earlier, when, engrossed in a grappling mating ritual and by means of a specially developed arm, a male had thrust into the mantle cavity of a female an elongated packet of sperm. A cap to each packet had opened later on, releasing sperm into the great cluster of eggs filling much of her body. Now females were standing on their heads across a gravelly area of the bottom, tentacles outstretched and bodies pointing straight up. Each deposited long fingerlike strands of transparent gelatinous material in which fertile eggs were evenly spaced. After attaching a few of these strands to a rock surface, a female made a vertical leap to a new position where she laid more eggs. In two weeks, the young would emerge to a planktonic life, each possessing only a few large color sacs on their translucent bodies, giving them a polka-dotted appearance that changed pattern and color constantly.

A few miles away the bonitos, no longer pursued by the white marlin, found a new source of food themselves that would demand all their speed. Close to the surface, they encountered a school of flying fish that immediately took alarm and fled upward. As each flying fish broke through the water film, it spread its enormous pectoral fins to serve as rigid wings, although the fish was still moving comparatively slowly. The fins elevated most of the body out of water, leaving only an elongated lobe of the tail fin submerged; this part of the fin sculled left and right so vigorously that the fish, all frictional drag from water now eliminated, dashed along at a greatly increased rate of speed. Within seconds its winglike fins took hold in the air and, with a last flip of its tail, the fish sailed high above the waves, its pelvic fins now also spread to give lift to the tail as well. Each air-borne flying fish glided over a hundred yards at a speed of thirty miles an hour, temporarily well out of a bonito's reach. The larger fish could keep pace with it in pure speed, but was unable to determine the direction taken by a flying fish during the time it was in the air. Many of the agile, specialized little fishes slipped into the water well out of reach of their pursuers.

While these rapid events took place, the fishes flashed by slower and smaller lives close to the surface. Microscopic plankton formed a dispersed blanket over the sea, but in diminished numbers from the warm, enriched bay. There were, however, new forms that were largely restricted to offshore waters by the great belts and currents circulating over the continental shelf. Elongated, conical sea snails— sea butterflies—swam with translucent blue "wings" in a crowded swarm. Each animal was equipped with reddish adhesive tentacles around a mouth armored with sharp teeth and hooks, using this impressive armament to feed upon other planktonic molluscs.

Among the lesser plankton were extremely small one-celled creatures that lived in long, pointed transparent cones, shaped like slender wineglasses without bases. From the opening, each animal thrust out a ringlet of microscopic hairs that beat in synchronized waves, creating a propeller-like action which sent its cone darting along, point foremost. These tintinnid protozoans had stubbier relatives close to shore, but the elongated ones were found primarily in the open sea. Other animals, not so active, were equally capable of remaining near the surface. Radiolarians, kin to amoebas, constructed intricately beautiful cases of glassy silica, each equipped with symmetrical spines that added to their surface area. The spines, together with an external layer of gas-filled bubbles that surrounded the outer silica shell, helped them to remain suspended in the water. Many of the radiolarians were globular, but some were elongate or conical, with spines pointing outward from one end. When they died, their empty siliceous shells drifted downward very slowly to become part of the varied sand composing the shelf.

Transparent larval snails and long-armed larval sea urchins added to the assemblage of other immature forms, making the plankton a youthful parade of the multitude of lives that swam in the surrounding waters or dwelt on the shelf below. Like the history of sea squirts in bay and ocean, immature animals of long ago, hardly different from those of the present, had often played significant roles in the emergence of new forms of life during evolutionary history. The adult copepods that swam in such abundance with their long curved antennae probably had evolved from the first larval stages of some ancient crab or shrimp, thus escaping the inherited fate of adult specialization. For many forms of life, these sudden, entirely new avenues opened by the ability of unspecialized larval stages to reproduce, meant an enormous proliferation into hitherto unfilled environmental possibilities. Copepods were the "insects" of the sea; not only was their abundance equivalent to that of insects which were the most numerous of all land creatures, but there was a similarity in origins, for insects had evolved from a generalized millipede larva and exploded into a million different specialties.

The copepods and many of the other active, swimming plankton were locked into what appeared to be windrows on the ocean surface. This was due in part to the refractive nature of waves which, as they passed by obliquely to the windrows, focused a flash of sunlight into the water below, restricting light-sensitive planktonic animals to specific bands. Depending upon the depth and temperature of the water, these tiny creatures were engaged in either a very energetic up-and-down activity while feeding on planktonic algae, or a much wider horizontal searching for more profitable grazing areas.

Algal plankton, nearly all diatoms, were the most prolific of the sea's plants, forming oceanic meadows of enormous magnitude. With five thousand diatoms to every half-inch cube from surface to a depth of one hundred feet or more, these tiny plant cells not only nourished all the animals of the sea through chains of grazer and predator, but contributed the bulk of oxygen to the earth's atmosphere—forests and prairies notwithstanding.

Ghostly chains of completely transparent animals hung linked together in the clear blue water of the outer shelf, their presence almost undetectable except for a small orange mass in each body and an occasional glint of sunlight on their barrel-shaped bodies. Each chain consisted of salps, a kind of free-living sea squirt, with the parent at one end budding a whole series of young from a long stalk. The chains swayed and spiraled as the cylindrical bodies of the salps contracted, forcing plankton-laden water through a capacious chamber in each individual. The total effect was graceful and haunting, for only the most acute vision could make out their presence and their extent.

Because of the proximity to the great current from the south, waters over the outer shelf often contained animals that had drifted up from the subtropics. In a place of meeting between two currents, dozens of heavy-bodied jellyfish were concentrated, their rose and brown bells rigid with firm jelly and bobbing in the swell. The pink and blue balloons of Portuguese men-of-war rode above the waves, trailing beneath them long tentacles which actually were three different kinds of highly specialized hydroids living in a floating colony. The longest individual hydroids served as food-gathering tentacles, for they were equipped with stinging cells and a retractile ability. Shorter, club-shaped hydroids, clustered against the underside of the gas-filled float, were purely defensive with their concentrated batteries of highly potent stinging cells. The last type of individual had a reproductive capacity: it set free tiny female jellyfish that later produced swimming larval forms, each able to live separately in the plankton while gradually secreting a bubble of gas that kept it close to the surface. As the gas bag grew, so did the number of individuals that budded from the first one, until the final colony possessed a float almost a foot long, with tentacles drifting down fifteen feet or more. On top of the float was a slightly canted crest that served as a sail, set to utilize prevailing wind conditions to counteract current drift in northern temperate seas. In the southern hemisphere, man-of-war sails were set in the opposite direction. The reversal of this one small feature in the two hemispheres was evidence of an age-long process at work: a natural selection during reproduction between those animals best able to contend successfully with the environment. Those with sails set the wrong way were carried far away and out of reach of others of their kind. Since they were unable to reproduce, their traits were lost to future generations.

Where the ocean currents passed one another over the outer shelf, the continent was all but invisible, marked chiefly by cumulus cloud patterns that capped updrafts over its warmer surface—the same water-bearing clouds that had figured so importantly in the river's early history.

Continental influences in the sea were fast disappearing. The life present was exclusively marine, except for one insect, an ocean-going water strider that traveled far out to sea, supported on the surface film by water-repellent feet in the same fashion as its inland relatives of pond and stream. Tree trunks floating down the river and out to sea were now largely destitute of whatever insects, spiders, or mites they had contained. Some logs had been in the water only two or three weeks, but already displayed colonies of stalked goose barnacles, boring isopod crustaceans, and shipworms which were really highly specialized clams with greatly reduced shells used only for boring purposes.

Floating debris from the land was one sign of the river's continuing influence and a slight discoloration of the water was another. Herring gulls, ringbill gulls, and a few other shore birds would fly out in fair weather to fish or rest on the surface, rising and falling in the great rhythmic ocean swells which were only beginning to feel the effects of the land. But the river was not done yet. It was still at work, hidden far beneath the surface of the open sea.

X

THE
FULL CIRCLE

The rising world of waters dark and deep. . . .
JOHN MILTON

A GRAY afternoon sky grew dim with the approach of strong winds and low, black clouds racing in from the sea, which was blue no longer. The sea was flecked with whitecaps across the dark rising waves. No ordinary storm was building: two hundred miles to the southeast a massive atmospheric disturbance approached hurricane strength. As the storm crept toward the continent, mighty waves generated by its winds grew in power, their rolling energy felt ever deeper until the bottom of the outer shelf began to stir nearly two hundred feet below the wild, wind-whipped surface. Slender, erect gorgonian corals swayed to and fro, and little swirls of sand rose whenever an unusually strong surge passed overhead, leaving ripple marks on the sandy bottom. This energy imparted new life to the remnants of the river, now all but lost in the wide sea.

Although the river as an entity was gone, its contribution of water, mixed thoroughly with sea water, and its suspended load of fine silt were still present. They would not be joined again, yet in a sense the river was reborn because of the storm. Some of the same water molecules that had fallen on mountain slopes and the same silt from distant eroding stream banks would be caught in a newly acceler- ated flow to a new destination.

The slope descending toward the ocean floor was very different from the near-level shelf that stretched one hundred miles from land, for it pitched down at a much steeper angle, dropping a foot lower for every ten in dis- tance. This incline continued for many miles until, at a depth of six thousand feet, the angle of descent decreased again and began to level out as it approached the abyssal plain—the floor of the ocean that lay twelve thousand feet beneath the surface.

The face of the slope was one of rugged grandeur, with great valleys and peaks serrating its surface that ran parallel to the continent. Rocky outcrops were exposed, alternat- ing with fine gray clay and sand compressed into soft layered sandstone. Because rock ledges were submerged continuations of sedimentary deposits originating far in- land on the coastal plain, occasional fresh-water springs emerged, limiting salt-loving marine populations within a brief radius. In similar fashion, continuous bubbling streams of trapped vapor rose from rock crevices, each shining expanding sphere filled with methane, a little carbon di- oxide, and traces of other gases.

In many regions, an extremely thin layer of mud covered lesser slope features, making it difficult to distinguish solid rock from loose material. The mud, composed of fine sedi- ment once held in the water, had been deposited by cur- rents coming off the shelf above and by extensive slippages of slope litter. Submarine landslides were not at all uncom- mon and were the primary means of preserving the angle of the slope. A million years ago they had occurred even more frequently from the massive deposits left by the retreating glaciers.

The slight incline of the continental rise in deeper water lacked many of the strong and irregular features of the higher slope, being quite smooth except where major channels descended through it from above. The contours of steep slope and gentle rise far below were not always typical of other offshore continental contours; a thousand miles to the south the drop, due to a fault in the crust long ago, was a virtual precipice, descending one mile for every two of horizontal distance. But these steep southern cliffs had no large river system behind them, fewer canyons nicking their edges, ánd no wide alluvial fans constituting their continental rise.

Sixty miles from the edge of the continental shoreline, in a region where the old river bed had broken up into a number of wandering channels passing through a delta-like deposit of sediment, a new depression, shallow and vague at first, ran straight out, cutting ever more deeply into the layered bottom. The farther it proceeded, the more it developed into a major feature of the shelf and the slope beyond, joined in its descent by many smaller gullies which emptied into it. The depression was the beginning of an enormous submarine canyon that was clearly related to the river and had no equal for a hundred miles in either direction. By the time it had cut two or three thousand feet into the shelf, it was a canyon of truly majestic proportions, with its steep V-shaped walls far exceeding anything on land in the eastern half of the continent. Like canyons far inland, this great submarine valley was fed by many lesser ones from a wide area of the shelf. Beyond the canyon, a wide apron of deposited soil spread far out across the abyssal plain as an extension of the continental rise, etched by shallow channels fanning out from the central gorge.

The beginning of the canyon had started more than a million years before, when the early swollen river had initially cut a shallow bed in the continental shelf, then largely exposed because of the lower sea level. River ero-

sion was not the sole cause of its existence, nor even a major
one any longer, for the deep canyon continued farther
down the slope than any ancient sea level had ever reached,
growing deeper all the time. Another more persistent agent
had been at work and it was this that continued to scour
out the canyon, leaving only a flat floor of sand between its
vertical walls; it was an agent that was also responsible for
the still-growing fan spreading across the abyss from the
continental rise.

Even though the force was not a direct continuation of
the river's flow, it was nevertheless related to the age-long
presence of the river. The river and its neighbors along
the coast had been active in carrying sediment from inland
to the shelf and slope, dropping it as their velocity was lost
in the sea. Some of the sediment had become consolidated
into layered rock, but much of the more recent material,
still lying on top of the shelf, was loose. As the power of
the growing storm waves far overhead reached the bottom
near the head of the canyon, sand and silt were disturbed,
lifted, and thrust down again in new locations. These particles were so agitated and had so little time to settle firmly
together that they became a part of the bottom water in
which they were mixed. The result was a turbid liquid mass
considerably heavier than ordinary sea water. It began to
slip into slight depressions, adding even more suspended
particles to its already heavy load. Where the slope of the
tributaries increased, the turbidity current gained in volume and momentum, until in only a matter of minutes the
entire middle canyon was filled with a raging torrent of
dense, liquefied sediment that scoured and pulled everything along in its path. It spilled over rocky outcrops in
cascades that resembled falls in the mountain brook, ran in
great sheets across the flat canyon floor, only to erupt in
riffles as it encountered round boulders that had been carried down in earlier flows. Some animals living at the bottom of the canyon were suffocated, others were whipped

up and rolled away, but a few survived. When it was all done, the floor would be very much as it had been before —only a little lower. Soon it would be brought back to its former level by regular minor contributions of sand from above.

The canyon ran straight down the slope where the current encountered only a few slight bends and irregular outcrops. Where obstructions were securely fixed, the current swept around them, cleaning their surfaces of silt and revealing layers of sandstone of great antiquity. Often the more porous rocks had been penetrated by boring animals; with weakened rims, parts of the steep walls collapsed as the current rushed by, adding to the material being transported downstream. Large sharp-edged boulders, smashed and torn from their resting places of millions of years, bounded down the canyon creating powerful compression waves that boomed through the sea as they careened from one rocky wall to the other. Fishes caught in the current were powerless to escape and soon were crushed or ripped into fragments. Sand dollars, sea cucumbers, coral, broken crabs, and all the multitude of lives from shallow water near the canyon head were carried by, as were ancient fossils, century-old water-logged tree trunks, nuts, twigs, clumps of grass still green from a stream bank, and glacial stones left eons ago by retreating ice sheets. Clouds of lesser sediment billowed up into the water, obscuring the canyon walls, later gently settling down to coat every surface with a fine film of silt.

The torrent roared through the deepest part of the canyon, four thousand feet below the sea surface, before emerging from the steep continental slope at over forty miles an hour. From there the mass of suspended material rushed out across the flat fan of the continental rise, confined to shallow channels with elevated rims. Four hundred miles from its point of origin and twelve thousand feet down, the great turbidity flow died in a whispering current, barely able to move the lightest silt particles. Like its innumerable predecessors, it had added a little to the wide fan, and filled the depths of the ocean a bit more with two or three feet of graded silt and sand. The very flatness of the abyssal plain over which the fan had spread was testimony to other flows over many millions of years which slowly filled what had once been a yawning trench. Beyond this flat floor, several hundred miles to the east, a ridge of submarine mountains rose, jagged and steep, creating a barrier between the even sedimentary plain and the rest of the wide ocean bottom which was craggy, pitted, stretched, and broken from stresses in the earth's crust when continents had wrenched apart long ago.

Over two hundred million years earlier, the continents now bordering the ocean had been one, but had separated as molten rock ascended into a great north-south rift. There fluid rock material, cooled into a solid crust underlying the ocean, was separated into two great plates that moved away from one another as new hot material ascended through the rift. Each crustal plate carried upon it the continents, their mountains and submerged continental shelves. In the deep flat plain between continental slope and distant submarine mountains, where the oldest fossils dated back only one hundred and fifty million years, the river reached the end of its final descent.

Turbidity currents had been descending the canyon for long ages, each one eroding a little more of its walls and bottom. The onset of these dense torrents had come about in a variety of ways, not always by deep storm waves. Even the efforts of a few boring clams could set off a current with the collapse of a section of canyon wall, but the most common cause of a submarine avalanche was the river itself: its slow accumulation of sediments near the head of the canyon continued until they became unstable and began to slip and cascade downward by their own weight. The river not only supplied the source of material for a turbidity current, but frequently the triggering action as well.

Even before the current of sand and silt had fully subsided, small fish darted in and out of the opaque stream, picking their fill of food. Crabs moved across the fan to the new deposit, which now was littered with organic debris, where they probed busily for soft worms and broken clams. As the silt settled and the water cleared, new ripple marks were evident across the soft bottom. Rocks that towered above the new flow were freshly scoured on the side that had faced the current, but mounds of silt had been deposited in their lee and sand clogged narrow crevices. The river still was at work in an endless task: the leveling of mountains and the filling of seas.

The canyon served as a gateway between the deep sea and the shallow waters of the shelf. Not many creatures of the coast could survive the tumultuous descent in a turbidity current, but a few did, as others had succeeded in doing in the past. The abyssal plains at the foot of the slope were populated in part by descendants of such involuntarily transported animals, as well as by a sprinkling of a few modern forms recently brought down and capable of surviving. Life had evolved first in shallow seas over three billion years ago and, for a while at least, the depths of the ocean, like the first barren continents, had been impenetrable and devoid of all living things. Invasions came about slowly, at times aided by sudden accidental transfers of the sort made possible by turbidity currents.

Passage through the canyon was not simply a downhill matter; the steep gorge allowed many creatures of the abyss and lower slopes an opportunity to ascend close to shore, where nutrient concentrations were not only plentiful, but accumulated in bulk near the head of the canyon until washed down once or twice a year. Just as animals of the intertidal shoreline had sought high, wet rock crevices as shelter from waves, many animals of the sea found in the

canyon a partial surcease from some of the rigors of existence. It was on the floor and rock walls of the canyon that
animals of the shallow sea could mingle with those of the
unlighted depths.

At the head of the canyon, life was most intense, for it
was here that the greatest concentrations of plankton and
organic matter were found. Copepod crustaceans were
especially active, each tiny animal filtering microscopic
plants from over two quarts of water every day. Copepod
numbers were such that no matter how rapidly diatoms
reproduced, their blooms were kept in check and it was
rare that dead plant cells floated along uneaten. Copepods
provided an essential and omnipresent link between surface
plants and larger animals as shimmering schools of anchovy
and other plankton-feeding fishes darted up from the shadowed walls, flashed in the subdued light, and ran down into
the gloom again, where they might be caught in turn. An
unseen predator caused a school of squids to leave hundreds
of puffs of ink slowly dissolving in the water as they darted
down into the shelter of the canyon.

In places the walls of the canyon consisted of overhanging rocks, creating caves and crevices heavily populated by
animals seeking cover. Although crabs, multiarmed starfish,
and grotesque, crawling fishes moved across ledge tops, the
undersides of granite and quartzite outcrops were fes-
tooned with thick, congested mats composed of a host of
attached creatures, often colorful but their colors muted or
lost in the dimness. Sessile forms flourished in the canyon
where plankton and organic detritus were regularly swept
along close to the walls. All that was required to feed
abundantly was a means of removing drifting materials
from the water—sieves, tentacles, traps, self-created currents, and other filtering devices. By growing on the near-
vertical walls or under ledges, these fixed animals had few
problems with the settling of suffocating silt; their populations were greater near the rim of the canyon than at the

bottom, where abrasion and scouring kept many of the rock faces bare.

In the shelter of overhanging rocks, the community of attaching life was more complex than in any exposed area on the shelf closer to shore. Because salinity and temperature remained largely unchanged, almost every kind of marine creature was present, making an even more varied association than that of the oyster beds in the bay. Here there were few dominant animals but a crowded mingling of types.

Inside many soft rock ledges, barely visible in their bored-out chambers, large angelwing and smaller boring clams slowly rotated, cutting and rasping away the sandstone. They were common toward the head of the canyon, but were rare at the lower levels. Other kinds of more exposed clams and mussels grew packed in the dense mat of life surrounding them. Lumpy sponges swathed the rock surfaces; often they were a dull gray-yellow but a few glowed vivid orange and red when an infrequent shaft of sunlight filtered down to illuminate the rock wall. Long-tentacled sea anemones reached down for passing prey, hanging relaxed and swaying in the current. On a much smaller scale, simple cylindrical animals similar to the sea anemones were an alternate generation of several of the familiar jellyfish of coastal waters. The larvae from such jellyfish had swum down, attached to an overhanging ledge, grown a little, and now were beginning to bud off tiny pulsating jellyfish again. This alternation between two very distinct generations had the advantage of wide dispersal and sexual reproduction one time and prolific nonsexual budding in a secure place during the following generation.

If the underside of a ledge could have been seen clearly, it would have been a place of singular beauty with its varied colors and hordes of peacock worms spreading rosettes of lacy tentacles into the dark water. When a rock crab passed by, clinging to the sandstone ceiling with sharply pointed toes, one peacock worm after the next popped back into its tube, flowerlike tentacles disappearing in an instant.

There were those encrusting animals that were not at all conspicuous: one kind of stalked sea squirt was numerous enough, but its tough tunic was so covered by debris and other smaller attached organisms that it was impossible to make out. Only its light-edged siphons moved slightly, responding with great sensitivity to the passing water. The sparse, shrublike growths of whip corals were another matter, for they reached two or three feet beyond the mat of low-lying animals, their tiny eight-tentacled polyps protruding a fraction of an inch beyond the slender reddish purple and orange branches, giving them a fuzzy and out-of-focus appearance.

A brown batfish, one of the most peculiar fishes in the entire ocean, crawled out upon a ledge from a crevice where it had been spending the daylight hours. Now that it was almost dark, the broad bumpy fish waddled out, shoving itself along by a pair of sharply angled muscular fins that resembled the hind legs of a frog. Each fin was raised in turn, so the fish actually walked along, its pointed head elevated and its tail dragging. While it could swim and frequently did at night to approach the top of the canyon, it was most at home walking along, looking for food with its froglike eyes. Just beneath its upturned snout was a small opening which contained a retracted tentacle; as the batfish crept up to a horned goby resting on a mat of yellow sponge, the short tentacle emerged and waggled furiously. When the smaller fish appeared interested and skipped closer, the batfish tensed, sprang forward and, with an inrushing gulp, swallowed the goby.

The sandy floor of the canyon also supported an extensive variety of animal life, even though it was affected periodically by rushing turbidity currents. Sea pens, relatives of whip coral, rose into the water from their sand anchorages, looking like stout, fleshy feathers, the branch-lets of which bore numerous small polyps. The thick, bulbous stalk of a sea pen had begun life as the first polyp of the colony, giving rise to all the smaller ones which collectively inhaled enough water to keep the parent in-flated and erect. Clams, tube worms, and burrowing snails were common here, but the most impressive population was that of sand dollars, their flat disks piled three or four deep, several hundred to the square yard. These relatives of starfish and sea urchins lacked the highly developed suction-cup tube feet of the others but were covered with short spines set close enough together to resemble a reddish-brown fur. Many of the sand dollars had been exposed by the recent current and now were in the process of burying themselves again. Working its thousands of spines in an organized fashion, each sand dollar began to tilt down a little, its forward spines throwing up a ridge of sand. When the mound was sufficiently high, the spines across the flattened undersurface of the animal pushed it forward, slicing into the heap of sand while spines on the upper, convex side helped to carry particles over the disk. In less than half an hour each sand dollar was once again beneath the sand, its presence suggested only by a faint circular outline.

The light had failed in the stormy sky far above, leaving a faint deep-blue glow in the water over the continental shelf and slope. The canyon with its steep walls was at last enveloped in darkness.

As the slope fell away, the dominant factor in the first thousand feet of water had been the light of day; but with the coming of night, surface waters became indistin-guishable from the depths many thousands of feet below. Because it was only at the surface that animals lived in the same environment as the producer plants upon which they ultimately depended, at night there was a general upward migration of mid-water predators and plankton feeders

from the regions of perpetual twilight in which they resided during the day. Life in the deep ocean relied upon
conditions at the surface in one way or another, even to the
extent of animals' passing their larval stages in well-lit
water before maturing and descending into the depths.
Whatever the cause, migrations up and down were appropriate and possible only for creatures living within the first
half-mile beneath the surface; those farther down in a zone
of complete darkness seldom ascended very far.

The plankton of mid-water depths, while nothing like
that of coastal waters, was nevertheless present and diverse.
A few kinds of heavy-walled diatoms and photosynthetic
flagellated cells found enough diffused light at a thousand
feet to maintain their activity and production of cell substance. They were vastly outnumbered by much smaller
types of olive-green plant cells encased in spherical calcar-
eous shells. At equivalent depths these coccolithophores
lived as other plant cells did; but at greater depths, as much
as ten thousand feet down, they were denied all light
energy and depended upon dissolved organic matter from
the rain of descending, decaying remains of surface life.
These minute plants composed three-quarters of the total
mid-water plankton and lived in such quantity that after
death their tiny shells sifted down toward the ocean floor
far below to add to its calcium deposits.

Animals feeding directly on the microscopic plants were
mostly crustaceans. No longer were copepods colorless and
translucent like those of the surface, but deep red or an
opaque black, with red and yellow oil droplets in their
limbs and bodies. Krill—elongated shrimps fed upon so
abundantly by whales and fishes in the lighted zone above
—here possessed large pink and red spots on their otherwise
transparent bodies. Their limbs were so finely bristled that
they formed efficient filter baskets even for the removal of
minute coccolithophores. Red amphipods with large reflecting eyes, still looking a little like river scuds, preyed

upon smaller crustaceans, as did rose-colored segmented worms that wriggled through the water with widespread paddlelike appendages extending from either side of each segment. Bright red arrow worms were the most abundant of the small predators, hanging quivering in the dark water before darting forward to grasp a copepod in their hooked jaws.

Red and brown bullet-shaped jellyfish throbbed along in layers around the six-thousand-foot level, capturing smaller animals by means of stinging cells on short, blunt tentacles. Still deeper, colonies of stinging hydroids related to the surface-dwelling man-of-war pumped along by means of pairs of swimming bells, each a specialized individual, while long trailing tendrils displayed feeding, digestive, and reproductive hydroids.

In the middle regions of the sea, halfway between the surface and the abyss floor, light and vision played important roles in the lives of animals, but in the gloom even the light of midday was barely able to separate the dark shadow of a fish from the surrounding water. The most conspicuous light source here was that of the animals themselves which had an extraordinarily wide range of light-producing organs.

The sea at these depths was not very heavily populated, although life forms were persistent enough to be scattered all through the region. Nearly all of them, whether fish, squid, shrimp, or jellyfish, were luminescent; the dark water sparkled and flashed with pinpoints, bars, or long series of living lights. Some lights were pale blue and violet, while others were whiter or yellowish. There were those which remained at a steady intensity, glowing or even producing beams that pierced the darkness. Other lights blinked on and off, or slowly grew in intensity and then faded away again. A region that had been dotted with tiny lights suddenly exploded in a rapid succession of great shining, billowing clouds which hid the escape maneuvers of several dozen squids. The bright fog was just as effective a screen as the dark sepia clouds produced by surface squids, except here the luminosity blinded and confused the pursuing predator before slowly diffusing and winking into extinction. That this was an efficient means of evasion was borne out by its adoption by several kinds of shrimps and fishes as well as by the squids.

A deep-sea squid possessed body colors that could never be seen: it might be pearly white, purple, sky blue, deep red, or a dark, rich blue. The movable pigment cells were arranged to open and close around luminescent organs, so light could flare out or be narrowed down to bright pinpricks. These squids generally lacked the powerful musculature of their surface relatives and were fragile creatures with soft bodies and veiled membranes between their shortened arms. Some of the closely related octopods were so frail they resembled translucent jellyfish as they pulsated through the water.

Weakly glowing living light emanated from some animals directly, but more efficient light was produced by luminescent bacteria held within glands or cells in an animal's body. Each bacterial cell produced a substance that was acted upon by an enzyme, but only in the presence of oxygen. A number of the animals holding such bacteria in a mutually beneficial relationship could temporarily shut down the flow of oxygen-carrying blood to the glands, thereby extinguishing the light for a while.

The cavities in which bacteria were kept were not always simple structures and some had become highly elaborated into efficient lanterns. The simplest ones had the bacteria shielded by a darkly pigmented cup, with light shining only out of the aperture. Others contained a reflective layer within the cup, doubling the light output. The most highly developed light organs possessed condensing lenses, color filters, and muscular diaphragms to enlarge or reduce the diameter of the beam. Light organs situated on the articulated eyestalks of deep-sea shrimps were movable: usually directed downward, they could be made to shine forward as well.

As the hours passed, pattern after pattern of living light moved into the field and out again. Rattail fish, their long belly glands gleaming, appeared swimming singly in the pursuit of lesser prey. Hatchet fish, with a whole series of large luminous organs shining beneath their deep, compressed silvery bodies, drifted in bright schools, while nearby lantern fish of a more normal shape displayed distinctive arrangements of lights along their sides, each pattern differing with the species. There were fishes with headlights directing intense beams outward; others that shone lights forward, while the majority illuminated the darkness beneath them.

Some fishes trailed luminescent barbels beneath their chins, sensitive organs that were tasseled, straight, or had lighted bulbs at their tips. A barbel responded to touch by an animal lured to its brightness, or to slight currents set up by the passage of another fish. There were fishes that had light organs within their mouths, attracting food with every inhalation of water. Lures were common among the fat, toothy deep-sea anglers which extended a long fin ray in front of their mouths. The slender stem was usually terminated by a bright, wriggling bait which appeared and disappeared according to the amount of oxygen the fish permitted to reach it. When another fish approached such a lure, the fin rod was drawn back toward the mouth until, with a sudden, yawning gulp, the victim was carried into the angler's enormous mouth. The mouths of anglers and most of the deep-sea fishes were equipped with very long, needle-sharp teeth which reduced the chance of a victim's escape, or the loss of a rare meal. As food was so difficult to find, many of these fishes had expansible stomachs which could contain prey considerably larger than the predator; the food was taken in through wide jaws hinged in a unique fashion to allow its passage.

The darkness was so nearly complete and the vastness of the environment so great, that recognition among the sparse populations was a matter of great importance if a particular kind of animal was to survive from one generation to the next. As two fishes passed in the dark, the lights of one triggered responsive flashes by the other. Only when the patterns were identical was there a possibility of drawing together for mating. While the arrangement of light distribution over an animal's body was vital to the seeking of a mate, many of the angler fish solved the problem in another way by having the male become parasitic and attached to the female during its whole existence, trailing beneath her like a tiny, accessory appendage.

The senses of fishes in mid-water were keen. A pressure-determining system, the same lateral line sense belonging to shallow-water fishes, was important here, as it enabled a fish to determine the least movement nearby. The eyes of fishes and other creatures in mid-depths usually were large and exceptionally well developed as long as there was even a trace of light from the surface. In much deeper water closer to the bottom of the abyss, eyes, luminous organs, and pressure senses diminished in size and importance. Although the eyes of mid-water fishes had lost their ability to perceive color, they were among the most specialized and sensitive eyes in the world. A rodlike cell in the retina of one of these organs could be stimulated by a single wave-particle of light, achieving the physical maximum of sensitivity. Each tiny nerve in the retina served many rod cells which were packed twenty million to a square millimeter; the result was an efficient form of energy amplification. In comparison to the eyes of surface dwellers, each eye had an enormous light-gathering lens and a pupil so wide that the iris could hardly be seen.

The eyes of some fishes were directed straight up and consisted of short cylinders rather than spheres. Inside each tubular eye were two different retinas, one for medium-distance vision and the other for use close up. Such tube-eyed fish gradually zeroed in on the vague shadows outlined above them by the infinitesimal amount of light filtering down through the sea, accurately coming close to their victims by means of their unique vision and pressure senses. Once underneath the other animal, there would be a quick swoop upward and the attack was complete.

Because of the extreme sensitivity of fishes' eyes in the twilight zone, there was the danger they might temporarily blind themselves when they turned on their main lights. Often the luminescing organs were placed so their beams shone downward from under the body or the tail, or they were situated toward the front of the head and separated from the eyes by a thick shield of dark pigment. There were even fishes with small, weak light organs inside their eyes which began to glow well before the main lights were turned on to condition the sensitive retinas chemically for the imminent glare.

The importance of making sounds in deep water was not less than that which produced the cacophony of shelf waters, only the noisemakers were fewer here and more widely spaced. Many of the calls resembled those of shallower regions—honks, cackles, and a deep drumming of muscles against gas-filled bladders. Now there were new sounds as well. Often they seemed to be used for establishing contact: a questioning beep, answered shortly by a deeper grunt from somewhere in the darkness. Far out in the black water, there suddenly came a shrill whistling squeal, immediately followed by deep bass notes of a prolonged, wrenching groan. The two contrapuntal cries continued at intervals, gradually dying away in the distance as whatever creature had created them slowly swam from the area.

Sounds had more function than merely identifying one

fish to another of the same kind, for they served as an echo-ranging device not unlike that of the bottlenose dolphin, although not so sophisticated. Once emitted by a fish, the sound bounced back from the bottom a mile beneath in a little over two seconds, enabling the animal to maintain a fairly even depth as it swam through the featureless darkness.

Where the depths plummeted down to the floor of the abyss, populations gradually diminished. Eyes grew less capable of forming images, but served only as light collectors, until they too became smaller, often vanishing under the cover of head scales. Light-producing organs were met less frequently, for the bottom was more of a two-dimensional world. Swimming close to it meant that sooner or later food and mates would be met.

The flat sandy floor of the abyss between the alluvial fan and the distant mountains constituted a monotonous environment that was swept by feeble currents. The deep trench which lay beneath the plain had been filling for millions of years and its present surface was not only a residuum of major contributions from land and continental shelf, but from space, the atmosphere, and the sea itself. Five million tons of dust from burned and pulverized meteorites fell to earth each year, much of it sifting down through the ocean. Incorporated in the fine sediments were ashes from distant volcanic eruptions and clay particles lifted from deserts on the other side of the earth, transported by high encircling belts of air before descending to the sea. Glacial rocks and gravel, carried out on rafts of ice, had descended the slope with vast quantities of mud from land. Layers of ooze consisting of the limy and siliceous remains of plankton still settled gently on the floor of the abyss, having built into thick layers during warmer and more favorable ages of the past. Some areas were cobbled mile after mile by large rounded rock nodules, the result of an extremely slow accretion of manganese and iron dissolved from the land around a glacial pebble, building intermittently at a rate of no more than a millimeter in a thousand years. When sediment covered the nodules, their growth stopped, but every so often a current would wash them clean and the slow transfer of the dissolved metallic elements from sea water to a solid base would commence again.

Fragments of old and distant lives lay scattered at random across the wide floor—the ear bones and vertebrae of long-dead whales, chunks of compressed, aged wood from far inland, and even windrows of stalks of marsh cordgrass, slow to decay upon the cold, lightless bottom.

Yet in this scene of apparent desolation there was life, especially where organic debris had collected. Bacteria carpeted the bottom and distinctive animals populated this single, most vast of the world's environments. One-half of the earth's area was that of the deep ocean floor, whether plain, trench, or rough and jagged fracture zone created as the continents had pulled apart. Surface waters were more extensive, but they differed widely in temperature and circulation patterns, whereas conditions in the abyss, polar or tropical, were much the same: cold, huge currents of slow circulation, a constant salinity, utterly dark, and well supplied with oxygen carried down by sinking water. Because of this constancy and the lack of isolation that is a necessity for the development of separate species, life on the ocean floor was much the same everywhere. Bottom animals had achieved a worldwide horizontal distribution, but were rigidly confined in their vertical movements.

Over the ages, as animals gradually invaded the depths and grew conditioned to the enormous pressure, they developed internal conditions that equalized those of the environment. Not all kinds of sea life had been able to penetrate so far down, for the pressure affected the formation of fats and proteins composing their cell structure, so only those creatures able to overcome problems of this sort had succeeded. All bottom dwellers were extremely sensitive to changes in depth and were clearly zoned to the slope, rise, and abyss floor according to their evolved abilities. Many of the simplest animals able to invade the abyss incorporated large amounts of water in their structure. The variety of fishes restricted to the greatest depths in no way approached that of the mid-water depths.

Life on the bottom required existence in a zone of transition between water and either soft sediment or rock. The simplest animals were cylindrical glass sponges, graceful woven baskets of glassy siliceous spicules, some of which were long and twisted in a ropy fashion, penetrating loose floor sediments to anchor the sponge in the gentle currents. There they sat, filtering organic particles from the water as it drifted past at a speed of only a mile a day. The slightest current set up by a sponge was sufficient to attract fishes which approached to nuzzle about in search of the blind crabs that usually crawled about the surface of a sponge colony. Tiny sea anemones clung to manganese nodules which also served as bases for stony and whip corals. Blue luminous sea pens and bulbous sea anemones burrowed into soft bottoms like their relatives farther up the slope. Tube-dwelling segmented worms attached to the nodules, to corals, and to any solid surface available to them, even if it was a slow-moving mollusc or starfish. When any of these animals produced eggs and larval stages, they did not rise far, but were swept along by deep currents to new places of attachment. Larval stages down here were short-lived and often did not exist at all, the parents retaining their developing young until they were ready to take up life on their own. This was especially true of the deep-sea fishes. A few bottom animals were incapable of reproducing at such extreme depths, their populations having to be renewed by breeding stocks halfway up the slope.

For food, creatures living on the abyssal floor depended to a large extent upon what fell from above, making use of

the energy contained within such organic matter. But the
particulate rain reaching the bottom to be acted upon by
bacteria was sparse, for it had passed through miles of water
where other bacteria and swimming filter-feeding animals
had consumed much of it. Occasionally, however, large
quantities did descend all the way. A few days before, a
fifty-ton finback whale, which had been ailing for several
weeks, died at the surface and sank to the bottom, only
partly destroyed by sharks which had been unable to keep
up with its plunging descent. Now it was the focal point for
scavengers on the bottom and for an explosive multiplica-
tion of deep-sea bacteria of decay. Crabs, fish, eels, worms,
starfish, brittle stars, and lesser animals approached the car-
cass, led by delicate chemical senses of odor perception to
the exact spot. Soon the huge body swarmed with scavengers and the region became one of intense activity. That
this had happened countless times before was borne out by
the irregular distribution of bacteria and bottom dwellers
across the abyssal plain. Wherever populations were concentrated, the bottom was pitted and furrowed with the
tracks of countless animals, but elsewhere it might be quite
smooth.

Because there was no dependable supply of food available to animals on the deep ocean floor and they had to
utilize whatever came their way, no one kind of animal had
its favorite prey, as was so often the case with creatures
close to the surface who had much to choose from. Logs,
fruits, and leaves from the river banks were not overlooked,
nor was animal excrement, the tough hornlike outer skeletons of crustaceans, or the finest particles of suspended
organic matter, too small to ever settle upon the bottom.
The last was filtered from the water by the most efficient of
straining devices in glass sponges, deep-sea barnacles, sea
lilies, and other sedentary animals. The sea lilies, relatives of
starfish and once populous in shallow seas many millions of
years ago, now found security only in the great depths

where they remained erect over the soft bottom, spreading their feathery arms into the current.

Great flattened sea cucumbers, the most abundant of the large bottom forms, inched sluglike through the sand and silt, using mouth tentacles to ram in and devour quantities of organic mud, the presence of each animal marked by a furrow and a tubular breathing device that extended above the substrate. Vase-shaped sea urchins, no longer resembling in the least the spherical pincushion types of shallow water, lay under the bottom sediment, almost all vestiges of their former radial symmetry now lost. There were other mud eaters on the bottom and a great many filter feeders, but both were greatly outnumbered by predatory forms of life.

Carnivorous clams and snails had a limited mobility in the muddy clay. Crabs, flattened lobsters, lesser crustaceans, and sea spiders were much more active, all rose-colored or vivid red, heavily armored, and all walking on widespread stiltlike legs that kept them high off the bottom. The length of their legs made many of these animals appear far larger than their shallow water relatives; the sea spiders, with a spread of a foot or more, were giants of their kind. Hermit crabs, often occupying deep-sea tooth shells as their coastal relatives had used snail shells, were almost hidden by the dark red-violet growths of corals covering their houses.

Some of the crabs were active enough to seize other crustaceans, squids, and an occasional fish, but lacking such opportunities, they would rip apart and eat glass sponges for the minute fraction of living substance they contained. Often the sponges yielded dividends, for they usually harbored corals, worms, and small shrimps within their basket-like bodies or attached to the outside. The eyeless crabs sought by fishes were very sluggish, passing their lives clinging to corals, stalked sea squirts, or sponges, where they also fell prey to their bigger and more energetic relatives.

The predatory fishes of the abyssal plain, rattails and big-headed brotulids being the most common, were also adapted to life on a remote bottom and differed from others of mid-water depths. Rattails, with their long, drawn-out bodies and projecting armored heads, rooted along in the soft bottom, exposing burrowing creatures in a fashion not unlike that of the river sturgeon. Once a victim was detected by a rattail's sensitive barbels, it was quickly sucked up into a mouth located far back under the fish's head. Furrows crisscrossed each other across the flat ocean floor, indications of the energetic gouging activities of these deep-sea fishes.

Across the lightless bottom a tripod fish, or stilt fin, jumped in quick and agile fashion, each time coming to rest on two enormously long fin rays emerging from either side of the front half of its body, and on one drawn-out ray extending the tail fin to a length equal to that of the entire body. When the fish rested in this position, elevated high above the bottom with its blind head angled upward, the stiff fins probed about in the mud for small crustaceans and worms that constituted its diet. As soon as prey was located by touch and uncovered by the fins, the fish's pressure sense detected its attempts at escape and it was quickly eaten.

Other fishes came and went across the wide abyssal plain, all solitary wanderers in search of food. Deep-sea gulper eels that could swallow animals many times their own size, sharks of a kind that had disappeared from surface waters eons ago, and robust, narrow-tailed chimaeras, relicts of a bygone race of sharklike fishes, swam slowly over the bottom or ascended hundreds of feet into the dark water when their senses told them prey was near. Every movement set up shock waves that traveled through the water only a little less than a mile a second, so the entire deep-water world was filled with radiating waves of pressure, the sources of which were constantly being sought by predatory fishes.

It would be a long time before water molecules recently entering the sea from the river would find their way to the abyss, but inevitably some would, just as those now present had at one time been a part of the sky and an ancient river. But the river's influence was felt, for the flat plain, constructed largely from turbidity currents, owed its existence to river action, no matter how remote the event had been. The vegetable debris littering the floor of the ocean beyond the fan-shaped rise was evidence of the river's more recent linkage of sea to the inner continent with its marshes and forested hills.

Two miles above the abyss, a pod of finback whales swam through the night toward the warm southern ocean a thousand miles away. The storm at the surface had passed over them and as the first warm rays of the morning sun filtered into the dark water filling it with a blue luminosity, they began feeding on schools of herring which blanketed the ocean for miles around. Whales, herring, and their planktonic food swam suspended in water that had been circulating in sky, land, sea, and in life itself for over three billion years. Some of the widely dispersed water molecules that had recently descended the river now were being incorporated in the tissues of organisms in complex chains of chemical events; some were swept aside by the vast gray hulks of the whales to remain as part of the ocean for a thousand years; and some began their quickened, twinkling dance at the surface where they would be drawn aloft into a waiting sky. The river was born again as it had been, and would be, every moment of earthly time.

XI

THE
INTRUDERS

*. . . Till taught by pain
Men really know not what good water's worth. . . .*
LORD BYRON

A BROWN man stood on the banks of the upper
river watching the ascent of salmon through the riffles, the
sweat on his bronzed skin drying in the cooling breeze
from across the water. He was soon joined by a small band
of his people who saw too the clean richness of the river
pouring through its green valley. With immediate agree-
ment that they should spend the next few days here, they
shrugged off the light loads they had been carrying and set
about making camp in a clearing shielded by several ancient
oaks.

Later, after fat salmon had been taken from embers of
the fire and devoured with relish, the man walked into the
shadows of the river forest and looked out over the shining
thread of water that captured the last light of a dimming
sky. What had seemed to be a fleeting thought entered his
mind, left, then returned again to remain fixed. Should they
stay here longer than his nomadic people had considered
before? Striking out across apparently endless mountains

and prairie was all they had ever known. He had no accurate way of recounting, but obscured in the tales they told around the fires at night was the history of the long journey of their ancestors from another continent, across a bridge of northern straits far to the west, and more slowly across the wide continent which was now their homeland. Their first arrival in this hemisphere had taken place many generations before, and for thousands of years still to come the dark-skinned people would penetrate every corner of the continent and its linked twin in the south, finally settling where they wished or when they could go no further.

The uncomplicated thoughts of the man as he stood looking out to the river presaged the future of his descendants, for the great tribe that grew from this one small band would remain in the river valley in the many years ahead until eventually they would be swept away by a human force they could neither understand nor contend against. Their settlements, established on headlands and river promontories, came to be of some permanence and consisted of long, arched bark huts surrounding central clearings. From there they went forth in dugout canoes to fish for what could be eaten at once, or could be preserved for exploratory trips inland along the many tributaries. They sought muskrats and waterfowl in the creeks, or collected oysters whose shells they discarded in deep pits. They were primarily men of peace, existing securely yet surrounded by larger warlike tribes who recognized the necessary skills the river dwellers could offer. Around their lodges, lost and sunk beneath the soil, were remnants of past barters with men from far away—stone tools and weapons that could not have been produced on the sandy coastal plain. A thousand miles inland, other men wore ornaments of shell, shaped by the river people from clams and oysters in the bay.

There were times when the river-bound men were conquered by invading tribes, but invariably they emerged free

again, often strengthened by converting a few of the war-
riors to their essentially peaceful and productive existence.
The men of the river lived with it in harmony, never
altering its constant sources of wealth.

After a great many generations had passed, a day came
when those who had settlements in the lower bay saw a
white-sailed vessel enter through the capes from the sea
beyond, a craft far larger than anything they had seen
before. During the few days the ship maneuvered through
the bay, they watched from the shore as pale-skinned men
swung weights from the ends of long lines and placed marks
upon sheets of parchment. When the visitors came ashore
for water, the river people faded back into the low forests
and were not seen.

Those who had watched these strange proceedings in
their youth were old when the pale men returned and
established a fortified log settlement close to the western
cape in the lower bay. Although the ways of the newcomers were strange, the river people met them without ani-
mosity and soon a limited trade was established. Toward
the end of the first year, one of the river chiefs suffered an
unwitting insult from a trader in the fort and, when he
replied in anger, was struck in the face. That evening the
brown men of the river, who had agreeably contended
with outsiders for thousands of years, concluded the stran-
gers had no right to their inherited land and went to war.
In two days, the stockade was overwhelmed; all its defend-
ers lay dead, bloody, and strewn about the inner compound. Their bodies were taken off and the fort and its
buildings burned, until all that was left was a charred area
which soon was covered by blowing sand and the slow
knitwork of sparse dune grasses. When more pale visitors
came the following year to supply their settlement, equita-
ble exchanges between the two peoples were no longer
possible; many of the river men and their families were
sought and slain before they could retreat upbay.

Within a few years, large sailing vessels were common sights in the bay, and the river people no longer dared to appear along the shore, but limited their activities to tidal creeks. New fortified towns sprang up, the creeks were penetrated, and long strips of coastal forest were cut to provide building materials and open land for cultivation of crops, many of which were new to the brown men, who watched from a distance or came softly at night to inspect the most recent invasions of their land. The slow, relentless pressure placed by one culture upon the other forced the river people inland away from the great body of water that had supported them for so long. As they retreated, occasionally resisting, many died from warfare, disease, and starvation, for larger surrounding tribes were also feeling the effects of the foreign invaders and had no wish to assimilate or aid the displaced river people.

The shoreline of the river soon took on a different appearance, with towns and shipyards and farm fields covering the fertile valley floor. Mills were established behind impoundments of tidal creeks and spilled their wastes to be flushed out by the tides. Small industries using water power gave way to larger ones as the centuries passed, until the upper estuary, by now a major shipping port, was lined solidly with enormous manufacturing plants that needed water for cooling or for shipment of products and the elimination of chemical wastes. The sewage from every bordering town or city emptied into the river. Great bridges sprang up until the river was crossed repeatedly by a webbing of concrete and steel.

By now, not a single part of the river's cycle, from clouds and rain, through brooks and the bay, to the deep sea, had escaped the effects of the new men. Plants, fish, birds, and mammals had been severely reduced by being collected, trapped, or destroyed; often they were replaced by organisms from abroad that flourished here without their original native checks. Ships brought creatures from

far away attached to their fouled hulls: animals and algae
that soon dropped off and took up residence in the bay and
estuary where they competed, often successfully, with
local forms.

The high atmosphere was accumulating waste gases
which retarded back-radiation from the earth, swaddling
the planet and causing its surface heat to build. Although
this was almost entirely a development of the past century,
in the next thirty years a quarter again as much carbon
dioxide would lie up there, forever out of reach of plants
that might consume it and oceans which could absorb it,
accelerating the increase of heat on the world beneath.

Other gases were collecting in dangerous quantities.
Long-lasting radioactive debris from atmospheric explo-
sions, such as strontium 90 and cesium 137, might circle the
earth in air currents for years before descending to be taken
up in bones and muscles of animals. Sulfur dioxide from the
combustion of oil and coal was injurious to life in its unaltered state, but could also be changed into tiny droplets of
sulfuric acid when it combined with water and oxygen in
the air. Nitric oxide and nitrogen dioxide, products of
internal combustion engines and industries, underwent a
series of reactions when exposed to ultraviolet from the
sun, resulting in smog that often hung over the river valley,
irritating eyes and damaging crops far beyond urban areas.

As the air blew in from the sea and crossed the coastal
plain on its way to the mountains, four hundred totally
different kinds of particles were present in the airstream,
many of them serving as nuclei for the condensation of
water droplets which would carry their chemical characteristics to earth. Only two dozen of these types of particles
might have been present in a world without man, and the
rest, whether powdered glass, rubber dust, paint spray,
lead, asbestos, scouring powder, rice hull dust, cement, cast-
iron grindings, acetate fibers, or any of the others, had become common ingredients of the air within the last one

hundred and fifty years of industrial production. The proliferation of industry was in itself a reflection of the increase in mankind, yet a cyclical process was at work, for
this same technology made it possible for men to survive in
ever-greater numbers. Their concentrated populations, sustained by continuing new developments in medicine, nutrition, and housing, made significant contributions to
atmospheric wastes. Persons crowded into two square
miles of an urban settlement produced as much carbon
dioxide as a very large industry. Man's favorite convey-
ance, the automobile, consumed as much oxygen as a thousand people and liberated even greater quantities of waste
gases, one of which, carbon monoxide, was so toxic it could
kill if sufficiently concentrated or at least have a debilitating
effect upon mental functioning.

Rain brought fertilizers, air-borne far from the site of
their original use, to the farthest corners of the continent,
enriching even distant, isolated lakes to such an extent that
some of their animals choked and disappeared in the resulting algal blooms, or suffocated when those same algae decomposed and bacteria consumed all the dissolved oxygen.
Yet this condition was slight compared to the huge
quantities of fertilizers swept into streams and lakes as water
ran across cultivated fields; the results of such pronounced
enrichment were then rapid and disastrous. Lakes far larger
than the ox-bow ponds of the valley became vast culture basins, supporting astronomical numbers of only a few kinds
of algae and tolerant worms and insects, to the exclusion of
all other life. Because of the runoff, the still water of
crescent ponds long ago had become opaque with dense
algal blooms, warm, and destitute of oxygen only a few feet
below the surface.

Once the clear cool headwaters of a stream supported
highly specialized life intolerant of altered conditions compared to the lower reaches of the river; they carried little in
the way of sediment. Most of the brooks and streams feed-

---

ing the river no longer had forested shores, but freshly plowed and unterraced farms, raw highway slopes, and construction sites bordering their banks. With little or no anchorage, silt, sand, and organic humus of the topsoil flooded off with every rain. At times, the wash from a hillside consisted mostly of mud carried in a vehicle of water, a soupy mass dumped into streams where the thick, impenetrable blanket of silt immediately reduced available food, destroyed spawning grounds of trout and other fishes and smothered their eggs. It severely reduced the ability of light to penetrate, so plants were few. The mud sometimes overflowed stream banks, entering towns and ruining homes, creating great expense to inhabitants who had on one occasion been obliged to remove as much as a million cubic yards of solidifying muck from a single community.

With stream and river banks cleared of vegetation, the whole process of developing meanders was accelerated, allowing streams far greater freedom to overflow and wander about the valley, inundating plowed lands and then removing their topsoil. The swift fish of clearwater streams, no longer able to feed or suffocated from clogged gills, were replaced by carp, imported from other continents, and by other mud-loving fishes of limited interest or importance to men and destructive of stream bottoms by their rooting habits.

The sediment, never stabilized, was carried down to wider, lower velocity areas of estuary and bay, where it settled over the oyster beds to such depth their community structure was altered and seriously reduced. Stream channels, lakes, and reservoirs had to be dredged repeatedly if they were to remain open. In lower rivers and estuaries throughout the country, it was necessary to dredge half a billion cubic yards each year to make them navigable, an enormous expense.

The river no longer flowed straight to the sea, but was

impounded repeatedly by dams which held its water for irrigation and power; one dam alone withheld over one hundred trillion gallons in a region where runoff from neighboring fields was excessive. In this and the other lakes now occupying the channel's course, siltation in the quiet water progressed rapidly, making each river lake shallower with lessened capacity. The quality of farms in the watershed area steadily diminished from the loss of topsoil, so additives had to be provided to sustain their yield. These additives—fertilizers to provide essential but lost plant nutrients and pesticides to control destructive insects, disease-causing fungi, and competitive weeds—soon left the areas for which they were meant and washed away into stream channels with every rain.

The adding of organic fertilizing compounds to streams in the river valley had been going on for over a century. Early pulp mills, converting forests to the material from which paper was made, emptied their sugary wastes directly into flowing water that carried the odorous and dangerous concentrations far away. Canneries, glue factories, laundries, dairies, tanneries, and other processing plants made similar contributions. The small towns springing up along the course of streams and the river discharged their domestic sewage, often raw and untreated, into the water, while livestock manure washed downhill into the same streams. Small quantities of these organic wastes at first served to mildly enrich the aquatic community, until festoons of algae streamed from every rock. Some of the fishes and lesser animals tolerated the impurities, but not for long as the intensity increased.

When the discharge of wastes had not been great in the early years of development, the river had rapidly diluted them and recovered its balanced populations in a natural purification process aided both by its turbulence and by its inhabitants. As the load of organic matter built up over the years, this cleansing ability diminished and even stopped in places, with little but bacteria and mats of gray fungus present to break down and digest the wastes as the foul stench of the river hung over mills and towns. Decomposers living on the bottom consumed all oxygen, so there were no fish or crustaceans, but only a few larval insects coming to the surface to breathe.

Downstream, past the towns and small industries, the river had still been gray and poisonous, but to a lesser degree. Pollutants grew more diluted and oxygen entered the water increasingly from carpets of blue-green algae and aeration in riffles at the surface; a few more animals appeared in the form of larval midges and filter-feeding protozoans that reduced the number of bacteria. As the stream sought its way downhill, it grew clearer and allowed sunlight to penetrate far enough to encourage the growth of green algae and tapegrass, both of which gave off increasing amounts of oxygen. Hardy fishes and more insects appeared as the stream continued to purify itself until it became clear in a complete recovery, supporting its normal varied community of organisms.

Many miles of stream flow were consumed in this cleansing process, especially where the river grew wider and more voluminous. When there were only a few settlements and mills along its course, long stretches of pure water intervened, but men increasingly developed the shoreline, so shortly after a section of the river began to purge itself of wastes, additional discharges entered and recovery was thwarted.

After technologies increased in scope, pesticides began entering the river, compounds for which there was no natural means of correction. Aquatic life was unable either to tolerate or to alter these stable new ingredients, and dilution, no matter how extensive, hardly lessened their cumulative effects. As the river progressed down its channel, it passed many kinds of farms and orchards, all of which used specific chemicals to control certain target pests such as weeds, fungi, mites, insects, roundworms, rodents, and others. Each chemical did its job effectively for the time being, although some of the pests might develop resistant strains in the future. The chemicals could not be aimed accurately at only one target, but invariably affected other organisms as well. Either directly or indirectly, the others sensitive to the agent were killed, rendered sterile, or driven away. In one region lying downstream of both farmland and an industrialized community, traces of nearly nine hundred basic chemicals were present in the river. It took only a few of them in the smallest concentrations—no more than three or four parts in a billion parts of water—to kill or immobilize typical river creatures. Crustaceans were among the first to be affected; far downstream in the estuary and bay where their populations had been enormous, their present depletion was widespread.

Often one pesticidal compound was far more abundant than others, having come from farms consisting of hundreds of acres of a single crop, sprayed or dusted against one specific organism to ensure large yields and unblemished produce. With so extensive an exposure, rain flushed off immense quantities of the chemical into the drainage system that led to the river.

Direct, early kills were only the first step, for as diatoms or crustaceans and other organisms diminished, so did those lives which ultimately fed upon them. Alternately, some of the smaller plankton were not themselves harmed by the chemicals present, despite absorbing them in some quantity. Well up the chain of predation, however, especially among fishes, the amounts of pesticides began to take their toll through a kind of biological magnification, for they became concentrated in the fat reserves of animals. They were thus preserved for transfer to larger predators or, when the animal which harbored them suddenly required fat-stored energy, they flooded its systems, killing it or affecting its reproductive metabolism. Predatory birds had been seen to tumble out of the sky in mid-flight and fish-eating grebes were found dead in the water. The osprey, another fish-eater, produced few viable eggs because of a failure to construct sturdy enough eggshells, an illness in liver function resulting from an accumulation of certain pesticides obtained at the end of its aquatic food chain. In the entire estuary and bay, where once osprey nests had crowned hundreds of trees, there were two nesting pairs left and only one had succeeded in raising young.

Recently millions of young fishes of one age had died all at once, killed by the absorption of the last of their yolk that contained chemicals which finally reached lethal proportions. Fish kills from different causes had been common during the past thirty or forty years, the earlier ones destroying two or three million river fish in a week's time, but later reaching totals many times that figure. In the last kills, two-thirds of the fish had died from the effects of industrial wastes, the remaining having succumbed to municipal sewage, acid mine wastes, and other pollutants.

The finding of a single mayfly was now an uncommon event; populations of frogs, snakes, and mammals associated with the river had declined because of chemical pesticides and other toxic substances flooding down the channel. And there were still more insidious consequences, for insoluble in water as many of these agents were, they continued unaltered out to sea, where they retained the capacity to impair life on the continental shelf and ultimately in the abyss.

Some of the ingredients in pesticides did not kill organisms outright, but induced slow-growing tumors or doubled the chances of mutation, breaking the genetic code in chromosomes at increased rates. The mutagenic pesticides often worked against the solution for which they were designed, especially among unwanted insects which, with their rapid reproduction, frequent generations, and large numbers, provided excellent possibilities for the emergence of new strains capable of resisting the specific action of the chemi-

cal. The unforeseen elimination of beneficial predatory insects, both on land and in water, was widespread and presented opportunities for newly immune pests to multiply
prodigiously. Predaceous diving beetles, once populous in
many of the small ox-bow ponds and slow streams, had
completely disappeared, leaving the stagnant water open to
great swarms of mosquito and gnat larvae.

As they were flushed down into the estuary and bay,
various chemicals not only severely cut into crustacean
plankton populations, but affected all manner of other creatures, their demise at first unnoticed by men. With less than
one part of chlorinated hydrocarbons in a million parts of
water, oysters were inhibited in growth, opened and closed
at the wrong times, and ceased their usual energetic filter
feeding. The levels of chemical pollutants in the bay were
not always dilute: when tons of pesticides were sprayed
regularly upon salt marshes to control mosquitoes, tens of
thousands of crabs also died with each application.

The river not only failed to reduce stable pesticide molecules to harmless ingredients or to get rid of them in some
other way, it was unable to deal with quite different, but
just as poisonous, substances set free in the water—metallic
compounds, phenols, salts and other dissolved solids, disease-
causing bacteria, fungi, and viruses. Concentrations of
salts, derivatives of crude oil, and ammonia compounds
could kill fishes at once, while their gills might become
damaged beyond recovery by acid wastes and dissolved
metals such as lead, zinc, and copper. Arsenic and selenium
compounds, usually present only in small quantities, were
stored gradually until they reached lethal concentrations.
Many fishes near the boundaries of polluted zones were
made ill by secondary infections which reduced their activity or their resistance, hence their ability to survive for
long.

Billowing clouds of foam in streams passing towns indicated the release of household detergents through sewers,

but even after foaming was brought under control by chemists, the bulk of such washing aids consisted not of easily decomposed soap, but of bulky, inexpensive phosphates that provided excessive amounts of unneeded fertilizer to the water. Other detergents, developed to break down quickly after use, produced poisons that were fatal to many aquatic organisms.

The conditions of pollution were accentuated in summertime when river flow was only one-sixtieth that of the spring peak, and when temperatures were most favorable to bacterial, algal, and fungal activity. The low rates of flow were further reduced when several hundred million gallons a day were diverted from the river to cities and farmlands outside the watershed.

In the tidal churning of the upper estuary, pollutants tended to accumulate no matter what the rate of discharge from cities and factories might be. In several places in the estuary and the river above, stretches of oxygenless water acted as impenetrable barriers to migrating fishes, crustaceans, or larval forms, isolating headwater streams from salmon, eels, alewife herrings, sturgeon, and other fishes. Even if a little oxygen should be present, the carbon dioxide content was often too high to allow exhalation of this gas by fishes.

When a rare shad attempted to penetrate the pollution barrier, it soon showed signs of distress from lack of oxygen: it became uncoordinated, swam rapidly and erratically before turning upside down to swim more feebly, its eyes and pressure sense system seriously damaged, finally sinking to the bottom where it rested against the sludge, belly up, gasping and dying. Such fish as were able to live long enough to be caught by a rare fisherman were unpalatable, tasting of petroleum or phenolic compounds.

The water off the great cities near the fall line at the head of the estuary was an opaque gray-black and gave off noxious fumes that hung over the valley for miles. The bottom was thick with sludge which constantly emitted gases from chemical and bacterial breakdown. On a still night, the whole river seemed to effervesce with foul bubbles bursting at the surface with gentle plops. The sulfide odor of rotten eggs coming from the water was more than many men could stand, and as these sulfides drifted in the air, they reacted with metals and discolored paints.

Lichens, once hardy pioneers on rock surfaces along the original shore, had ceased to exist and their gradual reappearance far inland was used by men as an indicator of the degree of atmospheric pollution, much of which emanated from the river or from industries located along its shores.

With over six million people in the largest river city, demands for water were heavy, as each person used approximately two hundred and fifty gallons a day. When the water was returned to the river, much of it carried high concentrations of sewage, still far from safe. These domestic wastes often resided in the area for weeks, oscillating back and forth in the tidal exchange of the estuary.

Heavy industries lining the river banks used many times more water than cities and poured into the river such a magnitude of waste no accounting could be made. Both population and industry were growing so fast that rates of water use would double in less than two decades and increase sixfold in the next forty years—only a theoretical matter at best, for the river would not be able to meet the demand.

Because various forms of heat exchange were essential in most of the industries bordering the river, water was used regularly for cooling. Often it returned to the river chemically unaltered, but raised several degrees in temperature, seemingly a minor physical change. Many organisms had very narrow temperature tolerances, so even a rise of two or three degrees was sufficient to erect barriers to migration of sensitive species, to halt or affect reproduction, and to reduce the success of larval distribution. Warmer water induced higher rates of metabolism in fishes and invertebrates, with dissolved oxygen being consumed more rapidly. At the same time warm water tended to drive off oxygen, which was rare enough anyway in the polluted river. Industrial water intakes usually lay close to the bottom of the channel, where the coolest water could be had, while discharge took place at the illuminated surface. The result was an increase in production of algae and microorganisms at the surface, which meant more organic material fell to the bottom, where bacterial decomposition produced further oxygen depletion.

Nuclear power plants, newcomers to the river, demanded extraordinary amounts of water for cooling purposes. Their exhaust water was pumped into basins or cooling towers where it lost much of its heat before returning to the river; nevertheless it took some heat with it, warming the river for several miles. The slight radioactivity it carried was of a feeble nature, but very long-lasting. As power plants of this type proliferated in rivers of the world, the accumulation of wastes such as tritium, a variation of hydrogen, would commence in the seas, building to unforeseen concentrations. Strontium 90 and cesium 137 from the reprocessing of reactor fuels were supposedly safely stored in tanks, but some of the containers were already leaking. Krypton 85, not stored, was released directly into the atmosphere. Unlike pesticides, radionuclides could not concentrate in animals' tissues beyond what the environment contained, so no immediate problem was anticipated. But what would the slowly flushing bay and inshore waters contain in half a century? Would the three dozen other radionuclides discharged into the river in lesser quantities have a long-term effect?

An industrial plant requiring water for cooling had to avoid sucking in the sludge and flotsam of the river; ordinary filters usually were of little effect. Once in a cooling system, the water was often heavily chlorinated to prevent clogging or sliming of tubes, filters, and basins; the chlorine eventually found its way into the river with the effluent from the plant, and was still sufficiently concentrated near discharge pipes to affect life in the immediate vicinity.

At times the river was submitted to only brief surges of hot water or toxic wastes, after which conditions apparently returned to normal, but not before the effects had been felt by sensitive organisms. Men were not always aware of the lethal nature of these transitory events, for by the time dead fish floated to the surface, there was no evidence of an environmental culprit.

In addition to the severely reduced life of the river or its total absence in some regions, the foul odors and black oils collecting on the surface, the channel was a trash receptacle. Its hidden bottom was a contaminating repository of the products of civilization, not quickly buried but repeatedly stirred and rearranged by the churning turbulence of great ships passing directly overhead, a thousand a month up and down the waterway. There had been a time when many of the containers cast into the river were constructed of disintegrating cardboard, metal which soon rusted apart, or wood that had floated away as it rotted. But many of the newer packages were aluminum, glass, rubber, and plastic, which would not oxidize, decompose, or float; they accumulated on the bottom in great heaps and thick layers, trapping silt and reducing the tidal flushing along the stream bed.

The river from the falls to the bay was septic, with curds, steam, foam, oils, floating sewage, viscous silt, lead, invisible acids such as phenols, other toxic chemicals, and multitudes of microorganisms filling its dark waters. This was the same river that four hundred years before had been described by the first men to enter it from the sea—the same who had been watched by the river people—as ". . . one of the finest, best and pleasantest rivers in the world," and two hundred years later inspired its new residents to write of its being "remarkably pure and light . . . everywhere pure, potable, perfectly salubrious. . . . In its extent, it is magnificent. In its form, it is beautiful . . . with a margin . . . ornamented with a fine fringe of shrubs and trees."

Upstream the trout were gone except where restocked by game commissions; the salmon could no longer reach its spawning grounds, due both to the mechanical blockage by dams and the chemical barriers of pollution. Seventy-five years ago, the towns bordering both estuary and bay had put forth fleets of shad and sturgeon fishing boats and large oyster schooners equipped for dredging. No one had seen a shad for two or three decades; the discovery of one sturgeon ten years ago had made headlines in a newspaper. Not a single shad fisherman was left, a four-million-dollar industry had vanished, and the last oyster dredge vessel in the lower bay was in the process of tying up for good after an unprofitable season where once a hundred thousand bushels of oysters had been harvested. Blue crabs were becoming increasingly scarce; clams carried the virus of hepatitis; the great oyster beds were almost dead, first from silting, then from pollution which affected their larval stages, and finally by an epidemic disease that destroyed all but a very few.

The astronomical numbers of menhaden that had filled the summer bay and offshore waters, and had loaded large seine-haulers to the gunwales each day, had almost vanished; the schools left were too small to bother with and the huge processing plants at the mouth of the bay had closed down. Whether their decline was due to disasters in the estuarine food chain or to excessive fishing practices that attempted to track them to their winter grounds, no one could say.

Excursion boats that had once plied the river with thousands of passengers, were tied rotting to city piers; the dozens of resorts they had visited along the river and bay

were closed permanently for reasons of health, marked
now only by toppling pilings that had once supported their
wharves. In the same location refinery piers stretched out
beside deep channels dredged to accommodate the largest
tankers.

Many of the river's problems were those of slow and
unforeseen growth of population and its demands; some
were the result of avarice; some were caused by compla-
cency that natural phenomena could cope indefinitely; and
some the result of accidents. Within the year, a tank truck
on a bordering highway had overturned, releasing intensely
concentrated pesticides into a ditch that ran into a stream
and then to the river; an antiquated city sewage disposal
system broke down for a few days; a ship containing phosphates collided with another, its cargo spilling directly into
the river. Shoreline refineries took precautions to prevent
oil spillage when tankers docked to be pumped full, but no
system was foolproof and a massive sheet of oil had floated
down the estuary, coating banks, marshes, a wildlife refuge,
and even recreational beaches far into the bay with heavy,
clinging oil.

Four decades ago, the bay had only begun to feel the
effects of the sick river that fed it, but now its vast populations were declining rapidly. Waters of the continental
shelf, which for years had received the contents of garbage
scows, were showing traces of the same pollution illnesses.
The sea waited beyond, but not untouched. Nearly half a
million substances now entered the river, many of them so
new and foreign that marine animals would have little
chance either to adapt or to contend with their gradually increasing concentrations. Far off the continental slope, an
occasional obsolete ship plunged through the gloom, carrying to the bottom its cargo of chemical poisons once designed for use against man, but now purposely discarded.
Sturdy containers of long-lasting radioactive waste had also
been dumped into the abyss with no assurance that the

activity would be ended by the time the containers themselves disintegrated in the sea. Fishes brought up from the deep ocean, or caught in seas far removed from an industrialized continent, had residues of pesticides in their flesh and other chemicals that had been in use only during the last two or three decades.

The depletion of life forms was nearly complete in long stretches of the river, but the seas were growing impoverished too. Fleets of powerful, capacious vessels used efficient devices, mechanical and electronic, to harvest the ocean of its animal populations, whether they rested upon the bottom, swam in mid-water, or schooled at the surface. Where once pods of two hundred finback whales had been common offshore, the sight of three or four individuals together was now cause for great excitement among those who sailed the coastal waters. The yield from the sea, once thought inexhaustible, was approaching its finite limits at a time when the illnesses of inland waters were contaminating the sea. Unpredicted and surprising, these events had come to the attention of man.

EPILOGUE

*All that is harmony for thee, O
Universe, is in harmony with me
as well.*

MARCUS AURELIUS

SOMEWHERE along the shores of primeval seas, over
three billion years ago, conditions that were the precursors
to life slowly became established. The ancient atmosphere,
unlike the one at present, was composed mostly of methane, ammonia, water, and hydrogen gases which were subjected both to strong ultraviolet radiation from the sun and
to lightning discharged almost continuously by the thick,
boiling storms that covered the earth. Complex molecular
substances, joined together by this energy, finally lay in
pools along warm coasts. Because there were no bacterial
decomposers or other consumers, it was the stability of the
molecules themselves that determined their permanence.
Some disappeared soon; others remained and grew more
concentrated; some added to their own structure from the
supply of lesser molecules available; and a few—the most
favored of all—were able to impress their own pattern of
organization upon newly created molecules that grew from
them. This was inheritance of the most basic sort. As they

proliferated, natural selection determined, still at a chemical level, which would be most successful and which would falter in the environment. Radiation and other sources of energy continued to alter their composition by random changes, until the warm, mineral-laden pools and sluggish streams contained a multitude of reproducing organic molecules, eventually composed of amino acids and nucleotides, the latter consisting of sugar, and phosphorus- and nitrogen-containing compounds. The most successful molecules concentrated environmental compounds around themselves in membrane-enclosed reservoirs—and became the first cells.

There was no single point in the history of organic chemical growth at which it could be said life appeared from non-life, for the whole possibility was a gift of an enormous span of time. Once there was a continuing line of reproducing cells, organized life had become established on the planet and would progress inexorably toward ultimate and unpredictable complexities of cell, multicelled organism, and community.

At first life was composed of large consumer molecules and simple cells building more of their own kind from the vast nourishing mineral soup available in the seas and the atmosphere, life that was totally dependent upon environmental ingredients for its structural continuity. But a time came when the supplies of intricate materials dwindled and growth and reproduction were affected. Those cells which demanded fewer and less complex nutrients were more fortunate, until the ultimate in efficient simplicity was achieved: the building of carbon dioxide and water molecules, plus a few other minerals, into the basic foodstuffs of life using sun energy and green chlorophyll as the means of construction. These first plants were bathed in sunlight because the original gases causing the thick overcast in the primeval atmosphere had finally been locked into life forms by earlier, primitive consumers. The other direction taken by life was precisely opposite to that of the plants: organisms which demanded more and more complex nutrients, not absorbed directly from the environment, but obtained by consuming the plants that had performed the initial work of selection and construction. So a reciprocal arrangement was worked out early in the earth's history as plants took in simple material from the planet's vast storehouse, constantly replenished by the decay of dead organisms, and inhaled the waste gases of animals; animals then breathed in the waste oxygen plants set free and devoured plant substance for the complex structural and energy-providing molecules they contained. Only the plants were versatile enough to exist on their own; the animals were totally dependent.

All of this was done either in water or through water as a carrier. Nothing could gain entry to a cell without the presence of water and all chemical mechanisms within a cell took place in a watery medium. The river circulated not only across the face of the earth and in the ocean deeps, but through life itself.

The source of energy for natural events on earth had always been the sun; the source of matter, the planet after it had formed within the solar system. The sun held the earth in place, impelled the atmosphere, oceans, rivers, and life in a closed system which required the use of the same units over and over again.

Ever since the universe began, the one fundamental process underlying all events had been that of a continuing creation from lesser levels of organization to more complex ones which emerged in unpredictable fashion—from subatomic particles to atoms, to molecules, to the giant molecules of life, to cells and their structured aggregates in the form of plants and animals. From hydrogen and oxygen gases one could no more predict water than a whale from a single cell, or the actions of an industrialized society from a Stone Age family.

Always there had been this drive toward complexity, this restless evolutionary striving of life to fit the environment. When a good fit was achieved, life endured and flourished; when the environment changed due to natural causes, life kept pace if it could, but if some forms disappeared, there never failed to be a pressure and replacement of life forms from the world's vast reservoirs. Life, the apex of creation, was persistent and filled every possible environmental niche; for over two billion years there had been few voids in the biosphere, that thin and fragile shell of life embracing the planet. Life, whose intrinsic form always reflects its function, was the continual recycling of earthly matter driven by the power of the sun.

Now the capriciousness of man altered the scheme and introduced a spreading degradation of nature. His creativity was nothing of the sort, but an interference and a destruction, bit by bit, of all that the world had built. His interruption and despoiling of the river's cyclical events and of life in water only meant harm to himself and to the world supporting him. Man drew nearer his own extinction on this small, isolated planet, pulling down with him many of the creatures that had sustained him during his short reign.

Yet man possesses the ultimate complexity denied all other life: he is the world—its physical and energetic components—at last aware of itself. His intellect is capable of perceiving the change and blight he has brought to the earth, as it is capable of halting and restoring some of the loss. He is aware there are too many of his kind; that by his very numbers he burdens the sky, earth, and its waters. As a conscious manifestation of the world's evolutionary progress, he is bound to the creation of new levels of complexity for survival. He still does not realize his fit in the world. If fear is a catalyst, he might reflect that in the history of life, being a misfit leads to extinction.

Many thousands of years after the first brown men had approached the river and lived with it, changing nothing, other men stood together on a long cement jetty used by a refinery. They looked toward the bay over the last wide salt marsh and debated its fate. Should it be preserved intact, ditched to drain away salt water and insect-breeding pools, sprayed, converted into a wildlife refuge, or filled and developed? At their backs lay over a hundred miles of heavily industrialized river shoreline and at their feet flowed the damaged, stinking river. Each man held firmly to his own belief in the importance of the river, yet their convictions differed and several were in almost irreconcilable conflict. In a little more than three centuries, these men had become as dependent upon the river as their vanished predecessors, and perhaps as much as the sturgeon and the osprey, but they had not yet learned to live in harmony with its age-old processes or its devastated populations. They had fulfilled an ancient destiny of their culture which admonished them to "be fruitful, and multiply, and replenish the earth, and subdue it: and have dominion over every living thing that moveth upon the earth," and to "create new heavens and a new earth: and the former shall not be remembered, nor come into mind." For nearly two thousand years they had followed this plan too well in different parts of the earth, increasing their numbers beyond reason and altering every region in which they settled. All but a few had come to regard the natural world as something less than mankind, whom they believed to be of divine origin and justified in exploitation of the environment. The advice they heeded had an element of disaster within it and their supremacy had not turned out well; their works had become spreading blemishes upon the film of life clinging to the earth. They were unable to see themselves as one part of a vast interrelating system, or to understand that dominion implied responsibility. In truth they had created a new earth with new skies above, unlike any condition the world had known before. The old world was nearly forgotten. Heavy economic pressures and immediate problems of a burgeoning human population had been thrust upon them and they still knew little of the intricacies of the world of air and water. By meeting and attempting to understand the wide turbid stream which passed before them with the sea as its ultimate destination, they took a hesitant step toward a more equable relationship with the great, dying river. But was there time?

I do not know much about gods; but I think that the river
Is a strong brown god—sullen, untamed and intractable,
Patient to some degree, at first recognized as a frontier;
Useful, untrustworthy, as a conveyor of commerce;
Then only a problem confronting the builder of bridges.
The problem once solved, the brown god is almost forgotten
By the dwellers in cities—ever, however, implacable,
Keeping his seasons and rages, destroyer, reminder
Of what men choose to forget. Unhonoured, unpropitiated
By worshippers of the machine, but waiting, watching and
    waiting.

T. S. ELIOT

About the Author

WILLIAM H. AMOS, a biologist, is Chairman of the Science Department at St. Andrew's School in Middletown, Delaware, and a Research Associate at the University of Delaware. He has been associated with the New York Zoological Society, American Institute of Biological Sciences and the Marine Biological Laboratory at Woods Hole, Massachusetts. In the academic year 1969–1970, he served as Visiting Scientist at the University of Hawaii. In addition to various monographs in aquatic and marine biology, he has written *Life of the Seashore* and *Life of the Pond*. He is a contributor to *National Geographic* and *Scientific American*, and his collection of photographs of marine and fresh-water subjects is one of the most extensive in existence.